John Gay

On Varicose Disease of the Lower Extremities and Its Allied Disorders

Skin Discoloration, Induration and Ulcer

John Gay

On Varicose Disease of the Lower Extremities and Its Allied Disorders
Skin Discoloration, Induration and Ulcer

ISBN/EAN: 9783337811273

Printed in Europe, USA, Canada, Australia, Japan

Cover: Foto ©berggeist007 / pixelio.de

More available books at **www.hansebooks.com**

WORKS BY THE SAME AUTHOR.

ON FEMORAL RUPTURE;

ITS ANATOMY, PHYSIOLOGY, AND SURGERY.

With Plates. 4to. *Price* 10s. 6d.

"If any surgeon, in the habit of operating on such cases, will try the method recommended by Mr. Gay, I feel assured that he will agree with me in considering that a vast improvement has been effected in the operation for Crural Hernia."—*Fergusson's "Practical Surgery."*

"The operation undertaken in this manner is little more than the taxis, with the addition of a superficial incision."—*Erichsen's "Science and Art of Surgery."*

A MEMOIR

ON

INDOLENT ULCERS AND THEIR TREATMENT.

12mo. *Price* 4s. 6d.

JOHN CHURCHILL AND SONS,
NEW BURLINGTON STREET.

THE ASPECTS OF MEDICAL SCIENCE.

𝔄𝔫 𝔒𝔯𝔞𝔱𝔦𝔬𝔫

DELIVERED BEFORE THE

MEDICAL SOCIETY OF LONDON IN 1860.

Price 1s. 6d.

LONDON: EFFINGHAM WILSON.

ON

VARICOSE DISEASE

OF THE

LOWER EXTREMITIES

AND ITS ALLIED DISORDERS:

SKIN DISCOLORATION,

INDURATION, AND ULCER:

BEING THE

LETTSOMIAN LECTURES

DELIVERED BEFORE THE

MEDICAL SOCIETY OF LONDON

IN 1867, BY

JOHN GAY, F.R.C.S.,

SURGEON TO THE GREAT NORTHERN HOSPITAL, CONSULTING SURGEON TO THE EARLSWOOD IDIOT ASYLUM, ETC., ETC.

"Disséquer en anatomie, faire des expériences en physiologie, suivre les maladies, et ouvrir les cadavres en médicine, c'est la une triple voie hors laquelle il ne peut y avoir d'anatomiste, de physiologiste, ni de médecin."

BRIQUET.

LONDON:
JOHN CHURCHILL AND SONS,
NEW BURLINGTON STREET.
1868.

LONDON:
PRINTED BY FIELD AND TUER,
136, MINORIES, E.C.

TO

CHARLES J. HARE, M.D. Cantab., F.R.C.P.,

AND

HENRY SMITH, Esq., F.R.C.S.,

PRESIDENTS

OF THE

MEDICAL SOCIETY OF LONDON

DURING THE SESSIONS 1866-67 & 1867-68,

THESE LECTURES ARE DEDICATED,

WITH EVERY SENTIMENT OF

ESTEEM,

BY THEIR FRIEND,

THE AUTHOR.

PREFACE.

I HAVE been induced to publish these Lectures at the advice of friends who, perhaps too indulgently, suggested that they might be of use in leading to renewed research into the diseases of which they treat; more especially as they indicate points of inquiry which of themselves might repay further investigation, and, in relation to Varicosity, yield results favourable to our hopes of bringing it more successfully, than it has hitherto been, within the resources of our Art. I have followed this advice the less reluctantly, inasmuch as I was unable to give, within the time prescribed for their delivery, more than abstracts of the matter which I had prepared, for the most part confined to inferences. In the Lectures as now published the facts upon which these inferences were based are also appended; and I shall be content if the one turn out to be but glimpses of truth, and the other its fragmentary transcripts. I cannot hope that these will realize more than this, on account of the many disadvantages with which I had to contend, and the little time I could spare for their preparation.

The Plates have obviously no pretensions, beyond being

copies on stone from sketches and notes made by myself from the several objects to which they refer, and which it was so difficult to render intelligible to another by mere description, that I resolved rather to give them simply as diagrams, than attempt anything more solely for the sake of artistic effect.

My acknowledgments are due to, amongst others, Dr. Richardson, Dr. Anstie, Dr. Reginald Southey, Mr. Savory, and Mr. W. Smith, for valuable aid whilst engaged in preparing the Lectures.

10, FINSBURY PLACE SOUTH,
August, 1868.

CONTENTS.

LECTURE I.

PAGES

INTRODUCTION.—ANATOMY AND PHYSIOLOGY OF THE SAPHENOUS SYSTEM.—INFERENCES IN RELATION TO VARICOSE DISEASE . . . . . . 1—53

LECTURE II.

MORBID ANATOMY. — SEATS OF OBSTRUCTION.— CURRENTS IN VARICOSE VEINS.—INFERENCES . 54—119

LECTURE III.

ETIOLOGY OF VARICOSE DISEASE; TREATMENT.— SKIN DISEASE.—DISCOLORATION.—INDURATION.— ULCER; TREATMENT . . . . 120—171

REFERENCES.

PLATE 1.

Fig.		Page
1.	The Internal Saphena and its Principal Branches	22
2.	The External Saphena	23
3.	Contraction of the last portion of Internal Saphena	28
4.	Channels of communication between the Saphenous and deep Venous System	33
5.	Valves in Saphenous Branches	30
6.	Communications between Venæ Comites	32

PLATE 2.

1.	Experiment to show the constantly Centripetal Current in Veins	33
2.	Dissection 4	65
3.	,, 7	67
4.	,, 8	69
5.	,, 10	71
6.	,, 11	73
7.	A Varicose Vein communicating with a Vena Comes	74
8.	Clot in Varicose Vein	74

PLATE 3.

1.	Saphenous Congestion in the Leg of a Dog, seven days after the destruction of a portion of the Femoral Vein	47
2.	Dissection 1—Tegumentary Varicosity	60
3.	Dilated Saphena	62
4.	Dissection 3	63
5.	Clubbed Clot and Clot with Tendrils	64
5.	I., II.—Dissection 13	76
6.	Embolism of Peroneal and Posterior Tibial Veins	64

PLATE 4.

1.	Dissection 6	66
2.	,, 14	78
2.	Fusiform Saphenous Varix	102
3.	Lamellated Clot in Venæ Comites	80
4.	Dissection 17	84
4.	Thrombosis of Posterior Tibial and Peroneal Venæ Comites	85
5.	Fibrous Degeneration of Adipose Tissue	156
6.	Dissection 20	88

References.

PLATE 5.

Fig.		Page
1.	Dissection 24...	93
ib.	x, x,—Lines of Incision in cases of Ulcer............................	169
2a.	Nodosities ; a, a, areolar tissue ; b, b, nodosities	102
2b.	Loculi ; or ampullæ, x, x, x, x..	102
3.	False Saphenous Varix..65,	106
4.	Peculiar Thrombus ;* a, sac ; b, fibrinous layer; c, clot ; d, band ...	111
5.	Varicose Veins, the Current in	114
6.	I., II.—Tracks of Varicose Veins	113
7, 8.	Tissue from inner wall of Loculus....................................	103
9.	Tissue from surface of Clot within recent lamellated coagulum in Varicose Vein	103
10.	Clot rolls, from Embolus	110†

* After these pages had gone to press I found that the thrombus here referred to, of which I had two specimens, had dissolved in some diluted methylated spirit, with the exception of some brick-dust coloured *débris*, which had fallen to the bottom of the bottle. This *débris* had become so changed that it yielded no specific results on microscopical examination.

† ¼ object-glass, Ross.

ERRATA.

Page 53, line 4, " by" instead of " to."
,, 63, line 6, " thoracic" instead of " thyracic."
,, 63, line 13, " Pl. III." instead of " Pl. II."
,, 85, line 16, it should be figure 4*.
,, 86, line 2, " Fig. III., fig. 4," should be omitted.

ON VARICOSE DISEASE

OF THE

LOWER EXTREMITIES;

AND ITS ALLIED DISORDERS,

SKIN INDURATION AND ULCER.

LECTURE I.

I HAVE first, Mr. President and Gentlemen, to acknowledge the distinction conferred upon me by being selected to the office which I have now the honour to fill in connection with our venerable Society,—that of its Lettsomian Lecturer.

On casting around for a subject on which to address you in this capacity, I met with my first and by no means a slight difficulty; for of the many that suggested themselves, I could scarcely find one that had not, in this restless and somewhat revolutionary age, been converted into a *specialism*, and comparatively exhausted through the labour bestowed upon it by its zealous appropriators.

The subject of Varicose Veins, although it cannot be said to have been entirely exempted from a like fate, did not, however, appear to me, on looking through its literature, to have profited thereby, in this country, so largely as other routine subjects. Points of great interest and importance in relation to the pathology of these veins, which have engaged the attention of our brethren on the Continent, and which may be said to be still *sub judice*, have received but comparatively little notice here; whilst some of the earliest and best contributions on the subject by our own countrymen appear, in the second-hand formularies in which they have reached us, to have lost much of their point and value; in all probability, from their having been submitted to a process of filtration, in the course of their transit, through varying strata of memories and monographs. Besides which, errors in relation to varicose disease, without any legitimate foundation in fact, and very little in fancy, have found their way into our modern works; and these have been so constantly repeated, both orally and typographically, as to have become dogmas of surgical faith; and all this because its investigation has been conducted too much in the library and too little in the deadhouse.

It therefore occurred to me that I could not more profitably occupy your time in listening to anything I could provide in the shape of lectures, or my own in preparing them, than by an attempt, from an appeal to the dead body, to confirm or refute the views that have been promulgated with respect to the subject of varicosity. But my preliminary difficulties did not end here. The little time at my disposal for such a task, when compared with its dimensions, and the difficulties that are thrown in the way of making dissections, or even post-mortem examinations in this country, made me *à priori* despair of anything like a fruitful issue of my undertaking, or such as could in any measure satisfy your expectations. *Unclaimed* bodies are buried by scores, that would, if utilized, give to anatomists here some of the advantages which are possessed by anatomists on the Continent; but the stolid prejudice against maltreatment of the dead body (would that it were as sternly carried out on behalf of the living!) evinced almost unexceptionably by our parochial authorities, forbids the chance of their being so profitably made use of.

However, through the obliging help of my friends Dr. Randall, of St. Mary's, and Mr. Fuller, his able coadjutor at the Marylebone Infirmary, and the late Mr. Clarke, of Shoreditch, I succeeded in obtaining opportunities of making a few dissections as well as some experiments, which I trust will be found to have contributed somewhat to the end proposed.

Of these opportunities I availed myself to the best of my time and ability; and to the observations which I recorded, and the inferences which I have been enabled satisfactorily to my own mind to arrive at, it is my purpose to call your attention.

If I should fail to convince you of the soundness of my conclusions, I think I shall at all events succeed in recording some observations that have the value of being transcripts, as far as they go, and it was in my power to render them, of their dead or living originals; and, what will be of still more consequence, in calling renewed attention to a disease which seriously affects both rich and poor :—the one by prohibiting beyond certain limits useful and invigorating exercise, the other by too often reducing the hard-working labourer, long before his day is done, to infirmity and the workhouse.

In extenuation of the many defects which will be found in these lectures, I must be allowed to plead a general paucity of those qualifications which are essential to the successful prosecution of an inquiry such as that upon which I have entered.

It is said of Mozart that he preferred to perform before the masters of his art rather than before his disciples. I venture to take courage from the practical sagacity of this great musician, assured that my efforts will find in my audience the indulgence I am too sure they will need.

The theme of these lectures will be Varicose Disease—principally that form known as varicose veins; and those other diseases of the leg which are most often found in association with it, and therefore supposed to be, in many instances at least, etiologically related to it,—viz., skin thickening, discoloration, and ulcer.

Varicose disease presents itself to us in the lower extremities in two somewhat distinct forms:—

1. In those figured patches of skin—"venosité" of Briquet—so often seen in fair and fat women; and which consist of clusters of diseased venous radicles. These tiny vessels are arranged *here* in the stellate, *there* in the arborescent form; whilst elsewhere they form themselves into a number of nearly parallel or converging lines, as though they tended to an unseen axial vein. They are all more or less beaded; and contain blood, sometimes of a *bright*, at others of a *dark* red colour. Sometimes their coats are stained; but more usually, as the blood is pressed out of them, so they disappear from view, until they become refilled. If these little vessels are followed, a single patch will be found in most instances to lead to a more or less curved and bloated vein of the size next to their own, and this to others gradually larger in succession until the trunk is reached. This form of varicosity is found in all parts, sometimes along the course of the saphenous trunks, but perhaps most frequently in parts remote from them.

2. As the familiar "varicose vein." A vein so diseased takes a more or less distorted course, and is often of considerable length, extending occasionally, without any very palpable interruption, from the upper part of the thigh to the ankle. Its general

outline is broken by several varieties of segmental configuration, from the merely wavy serpentiform or convolute to the severely tortuous. Its size and dermoid relations vary much; for at times it is small, and appears lazily to meander just below or scarcely above the level of the skin; whilst at others it stands out in bold and vigorous relief, almost entirely above its plane. Cases are recorded in which these veins have attained an enormous size, and in which the transit of the blood is accompanied with an arterial thrill. An instance is recorded by Mr. Colles,* in which the veins about the knee were said to have been quite as large as the vena cava, and others as large as the portal vein. "Every one of these veins had the thrill characteristic of an aneurismal varix, beyond which there was nothing to warrant the supposition that they communicated with arteries. It is said, however, that they scarcely impeded the man's walking or interfered with his usual occupation—that of a farm labourer." Other cases of the kind are reported; one by Mr. Goss, to which I shall have occasion to refer hereafter.

In the thigh, varicosity is met with in one or two, rarely more than three or four, of the principal branches of the saphena, or their secondary tributaries; and in the leg their number, even in severe cases, is not greater, although, from their proximity to each other, they often appear to be comparatively numerous. About the ankle, however, they often exist in such numbers as to form a wide-spread and inextricable plexus; and, in the foot, a series of parallel veins, which creep over its edge and front towards their larger receptacles.

The disease is in most instances confined to the primary

* Colles's Lectures, by M'Coy.

saphenous branches, and does not affect their next tributaries. In others, on the contrary, the morbid change extends itself from the primary branches throughout their principal ramifications. These distinctions mark two different forms of the disease—viz., the *sthenic* and *asthenic*: the first being met with in robust persons, and without any corresponding disease in other parts of the venous system, as in the hæmorrhoidal or spermatic veins; whilst the latter is found in weak persons, and is often associated with disease in those veins, with *arcus senilis*,* and other indications of general laxity or degeneration of fibre.

The veins themselves present a variety of objective features, according to the precise stage of disease that each vein or particular portion may have reached. *Here* the vessel is firm, elastic, and with little superficial unevenness of any kind; whilst *there* it is uneven, and in parts so attenuated, that vein-wall and skin together form a very thin barrier between this world and the next,—so thin that the latter is sometimes reached by a slight strain of the body; and its track presents a mottled aspect, blue and white.

The vein, too, gradually channels out the skin, so that after a time the finger, passed along its track, sinks into a furrow; and in advanced forms of the disease it moreover meets with a series of more or less sharp nodosities, each of which occurs at the bend of a convolution, with other less pronounced indurations or thickenings, which are usually associated with the vein pouches. These nodosities and thickenings are usually but erroneously attributed to diseased valves, phlebolithes, &c. Then, with varicose veins the skin, in the lower part of the leg more especially, often becomes hard, unyielding, and of a peculiar

* See Canton on "Arcus Senilis."

brown or bronze colour; and ulcers form, with or without the change just indicated. The common co-existence of these separate forms of disease in the same part of the leg has led to a belief in their pathogenetical association; and the "varicose ulcer" has become an axiom. It is my purpose to examine into this alleged relationship, and to assign to each its immediate causes upon a more satisfactory basis than that of mere coincidence.

The ancients appear to have known very little of varicose disease beyond what was revealed to them by its external aspects, and the issues of their modes of treating it. The Greeks thought a varicose vein resembled a cherry, and so they called it " κιρσος ;" whilst the Romans, from its uneven outline, called it "varix." It is somewhat strange that, to the imaginative minds of the fathers of our art, its snake or worm-like appearance had not suggested a term of appellation derived from the resources of their ophidian or annelidan nomenclature. It is, however, a matter of little moment.

There should be very good reasons for changing names. The substitution of an heraldic surname, for instance, for one that is offensive seems intelligible enough, but not so that of "Phlebeurisma," as devised by Ploucquet, or "Phlebecstasis" by Alibert, for Varix, for they give very little if any better idea of the disease itself; and although these classical synonyms found considerable favour at one time, they have fallen into comparative disuse; and we have become so familiar with the term "Varicose," and with it our ideas of the disease with which it is customarily linked are so closely identified, that I have determined to retain its use in the course of these lectures. But instead of applying this term to that especial form of disease with which it has hitherto been associated,—viz., to that which is met with in certain superficial veins, I propose to use it *generically*, and to include under " Varicose Disease,"—

1. *Varix*, or partial and limited dilatation of a vein ; and,—

2. *Varicosity*, or its *general* dilatation, with those other morbid changes to which the well-known varicose vein is prone.

This distinction is necessary, not only because the two forms of disease differ materially ; but also because without it, the allusions to the veins that are met with in some of the older and abler writers on this subject cannot be clearly comprehended.

One of the earliest anatomical definitions is that given by Paulus Ægineta :—" Venæ dilatatio, aliquando in temporibus, aliquando in ventris parte, sub umbilico, nonnunquam etiam circa testiculos, sed maxime in cruribus ; "—a definition that was accepted by Severinus, Baillie, Heister, Sharpe ; and has been, indeed, by later writers so long as varicose disorder was supposed to be fully expressed by its external indications.

Fifty years ago Mr. Hodgson, in his admirable treatise on the " Diseases of Arteries and Veins," invited more earnest attention to the subject of varicose disease than had before been given to it. He investigated it in the dead body ; and drew distinctions between *uniform* and *partial* dilatation of a varicose vein which have not met with due appreciation by modern pathologists.

Mr. Hodgson showed that varicosity is, for the most part, confined to the superficial veins ; but I do not think that in this remark he meant to include the saphenous trunks ; for in reference to the condition of the external saphena, he states that "it is dilated throughout its whole extent, but in some places it is *expanded* into tumours," a description answering to a species of dilatation which that vessel often undergoes;—*a varix*, but not a *varicose* dilatation in the sense in which this term is applicable to its branches. In Case 51 Mr. Hodgson says "there was a cluster of enlarged (*varicose*) veins on the side of the calf of the leg, and a varix as large as a pigeon's egg in the course of the saphena below the knee." In no one case does Mr. Hodgson distinctly state that he found varicosity—the ordinary affection of varicose veins—in the saphenous trunks.

Mr. Hodgson, in adopting the method of treating varicose veins by ligature of the saphena, expressly states that it was not for varicosity of that vein; and in his remarks on Case 53 he says, "The veins of the *foot and leg* were varicose, and as it was most probable that the intractable state of the ulcer depended upon this condition of the veins, the *vena saphena major* below the knee was tied."

I must skip over a considerable period and several works, and refer to an admirable practical treatise on varicose veins by Sir Benjamin Brodie in 1841. He described a varicose vein as "preternaturally dilated without the dilatation being instituted to answer any good purpose in the animal economy."

Sir Benjamin was the first in this country to make a clear distinction between *tortuosity* of a vein from obstruction of its trunk, which he says is a "*healthy* state of the veins," and *varicosity*, which is essentially a diseased condition; and supports the view current throughout Mr. Hodgson's work, that it is the *superficial* veins that become varicose, especially *the branches* of the *vena saphena major*, and *sometimes* of the *vena saphena posterior*, the small veins becoming first affected; "but," continues Sir Benjamin, "as the disease proceeds, it extends *to the trunk* of the *vena saphena major*, which *becomes knotted all the way up to the groin*. Sometimes this vein is as large as your finger, and assumes a *knotted* appearance; the knotting being occasioned by the lengthening and consequent twisting of the vein, as it were, upon itself."

Throughout this paper there is presumptive evidence that it was written from observations made *principally* on the living subject; and I can only reconcile Sir Benjamin's account of the *knotted* and, "as it were, twisted" condition of the internal saphenous trunk with the results of Mr. Hodgson's observations, as well as with those of more modern research, by assuming that it either must have been a very *exceptional instance*, or that it must have been a branch of that vessel, and not the vessel itself, which was found to have become so contorted. I may here anticipate a remark that the internal saphenous vein, especially its femoral portion, usually lies so deep, that when excessively dilated, it is not always clearly distinguishable to the touch; and that a branch which often runs along on a higher plane, but on the same course, is constantly misapprehended for it, especially when varicose. In my dissections I have never been quite sure that such a vein has not been the saphenous until I have proved it by further search. The deep-seated veins, according to Sir Benjamin, never become varicose, because there is the pressure of other organs upon them, which prevents their dilatation.

In the "Cyclopædia of Anatomy and Physiology" Mr. Salter says that varicosity is usually "confined to one branch of the internal saphena; the external saphena may, however, be affected, or several branches of each."

In 1852, continuing this cursory review of what has been done in this country, Mr. Nunn published an interesting work on "Varicose Veins and Ulcers." Mr. Nunn gives an account of a dissection made of "a case of old standing varicose ulcer" by Mr. Bridgewater. He found the saphenous vein above the knee simply dilated and thickened. The tortuosities were below the knee; but whether they belonged to the trunk vein or only to its branches is not quite clear, for it is stated that "below the ulcer the vein is not tortuous, but thickened, and its lining membrane roughened; whilst at the seat of the ulcer itself it lies in a bony channel, formed by the thickening of the periosteum and the deposit of bony matter on either side." By implication, however, I think they were the *branches*, and not the *saphenous* veins themselves. The state of the deep veins is not given.

On the other hand, Mr. Chapman, in his work on "Varicose Veins," in common with several other writers, uses the term varix to denote " dilatation of the trunk and branches of the saphena; " thereby meaning, I apprehend, that these several vessels are equally prone to the same form of disease, viz., varix; and that this disease is identical with that of a vein commonly known as " varicose."

Mr. Callender, in an able article in " Holmes' Surgery," appears also to countenance the idea of *saphenous* varicosity; for he says, " Varices do not invariably commence in the trunk of the saphena or in that of any superficial vein;" adding this most important remark in confirmation of a like statement by Briquet, that " as far as superficial varices are concerned, wherever the intra-muscular veins pass into the subcutaneous, there varix is first noticed;"—assigning as the reason, that "the valves obstruct below, and the column of obstructed, slow-moving blood resists above."

In the last edition of his treatise on "Diseases of the Veins, and Hæmorrhoidal Tumours," Mr. Lee does not affix varicosity to any particular veins; but in one case he speaks of "varicose enlargement of the external saphenous and tributary veins of the left leg."

I must now take a retrospect of what the French school has contributed towards our knowledge of this disease, keeping as much as possible to its anatomical relations.

Boyer defined varicosity as "les tumeurs noueuses et inégales; formées par la dilatation, contre nature et permanente, des veines souscutanées," and observed that it affected principally the superficial veins of the leg, the deep veins being rendered almost immune through their muscular and internal connections.

One of the most remarkable monographs on this subject extant was contributed by M. Briquet, and is published amongst the "Thèses de l'École de Medicine" for the year 1824. "All the veins of the body," says M. Briquet, "may become varicose, but the disease affects principally the veins of the leg; and these in the following order:—1, the internal, and 2, the external saphenous."

Although M. Briquet gives apparently to the *saphenous vessels* themselves the greater share in varicose disease, yet on studying his paper attentively it will be clear that he means their *branches* rather than their *trunks;* for in his pathological account of the disorder he divides the various phases of the disease into—1, simple dilatation; 2, uniform dilatation with thickening, which he says *is especially seen in saphenous trunks, seldom in the branches.* "*The walls are thickened often like those of an artery; and the vessel is usually straight, with occasionally the least inflexion;*" and 3, unequal dilatation with thickening or thinning of the walls of the vessel, which he says affects the saphena below the thigh and the principal branches in the leg. "Les veines profondes," adds this excellent observer, " sont tantôt dans l'état sain ; d'autres fois je les ai trouvées fort épaisses, devenues presque semblables à l'artère qu'elles côtoyaient. À l'endroit où la phlébecstasie est le plus prononcée, il y a des communications très-larges avec les veines profondes, que sont larges à l'endroit d'où part la branche anastomotique; mais qu'ils reprennent leur calibre. Je suis persuadé que des injections faites dans l'artère iraient très-librement dans les veines."

Attention had just been called to the state of the deep veins in cases of superficial varicosity. M. Briquet had seen the case of one at the Salpêtrière, in which, although the external parts of the legs and thighs were covered with large veins, varicose to their smallest ramifications, *the deep parts of the limbs were pale and free from all signs of disease;* from which M. Briquet concludes that "these cases are far more frequent than would appear from the works of those who appear to regard this affection as a *general* disease."

M. Broca, in order to test the validity of M. Briquet's statements, examined a case of varicose disease, and found the deep veins coincidently affected. M. Denucé found another in which, on the contrary, the deep veins to their muscular tributaries were dilated to a high degree, whilst the superficial veins were exempt from disease. Other cases confirmatory of the same fact were furnished by Deville, Fouché, and Dumay.

In Cruveilhier's splendid work, varicosity is illustrated by a figure of a saphenous vein, which, with some tortuous tributaries, has a loop branch which is decidedly varicose. The saphenous vein itself is perhaps somewhat thickened; but it is free from dilatation, with the exception of a single ampullary dilatation, or varix. It is also thrombose. In another figure the terminal portions of some saphenous branches are slightly varicose, but the trunks themselves are not implicated.

In 1855 the investigation of this subject was undertaken by M. Verneuil, and pursued at the cost of immense labour, with great ability, and all those resources for conducting such inquiries which our French brethren so pre-eminently enjoy. M. Verneuil made twenty-one dissections, and published their results in three articles,—one contributed to the *Gazette Médicale de Paris* for 1855, p. 524; another to the *Gazette Hebdomadaire* for the same year, p. 811; and the third to the same journal for 1861, p. 428. M. Verneuil concludes from his dissections,—

" 1. That whenever idiopathic (spontanées) superficial varices exist in the lower limb, the corresponding deep veins are also varicose."

" 2. The correlative position is not, however, proved; for the intra- and inter-muscular veins have been found to be dilated without any affection of the superficial veins: but when the *first* set are alone dilated, it is almost certain that after a time the second will enlarge, become serpentiform, and visible through the skin."

" 3. Phlebecstasy in the lower limbs, does not *begin* in the subcutaneous veins, neither the internal saphenous nor any other; but on the contrary, in the deep veins in general, most frequently in those of the calf of the leg. These vessels first dilate, then the valves become inefficient, and the lesions extend to the second and third rank subcutaneous veins."

In the second article (*Gazette Hebdomadaire*) M. Verneuil arrives at the following further conclusions. "That idiopathic phlebecstasy never commences in the trunk of the internal saphenous vein, but in its secondary and anastomosing branches. The saphena itself often remains healthy, but more frequently it wastes at least in the leg, even when the entire member is covered with venous dilatations.

" Far from being rare, deep varices are more common than subcutaneous varices."

From these observations M. Verneuil draws the following inferences:—

" 1. That the primitive seat of phlebecstasy resides in the deep veins. These first suffer dilatation for reasons which anatomy and physiology render imperative; and from these it passes to the subcutaneous veins."

"This extension takes place by various anatomical channels which exist between the superficial and deep veins."

"I affirm," says M. Verneuil, "that if you find in any part of the limb a spot, ever so limited, in which the superficial veins are serpentine, if you trace them with care, you will find that they communicate by large tracks with the deep in*tra*- or in*ter*-muscular branches; and that any exceptions to this rule are readily explained."

I have already referred to Mr. Callender's paper in " Holmes' Surgery." I must, however, return to it, as Mr. Callender has, I know, taken great pains to determine the two points raised by Briquet and Verneuil, and upon which these authors are at issue.

Whilst Callender confirms the statement of Briquet that "at the spot were varices are most common, large trunks communicate through the muscles with the deep veins;" adding, that "the deep veins are as commonly diseased as the superficial vessels, the disease being most advanced where the intra-muscular veins empty their blood into either set;" he has failed in the course of his dissections to discover evidence in support of the constancy of *intra-muscular varix*, as insisted upon by Verneuil.

From this cursory review, it is clear that a considerable amount of hard and honest labour has been bestowed upon the investigation of varicose disease; and with the following as perhaps the most important results;—first, in relation to the especial seats of varicose disease; and secondly, to the implication of other departments of the venous system, when this disease is met with in the superficial veins.

1. It is asserted by some that varicosity affects the *trunk*, as well as the branches of the saphenous veins promiscuously. The question has, however, been left somewhat at issue; for although it is more than probable that Mr. Hodgson did not countenance that view, and that it has not been supported by M. Verneuil, it has yet received the support of Boyer and Briquet, and in our days it is so commonly accepted that a varicose saphena is almost a surgical canon.

2. According to Briquet and Callender, with superficial varicosity "at the spot where varices are most common, large trunks communicate through the muscles with the deep veins;" and,—

3. That, according to Verneuil, superficial varicosity invariably commences in the " deep veins," and frequently, if not constantly, in the intra-muscular veins, whence it is propagated to the former by the numerous anastomoses which exist between the two sets. To this conclusion Briquet's observation do not tend; neither has Callender, although, as we have just seen, he testifies to the constant communications between the superficial and deep veins, met with evidence in support of the constant implication of the *intra-muscular* branches.

I shall now give a short summary of the principal theories as to the etiology of varicose disorder according to various authorities.

1. *Mr. Hodgson* referred varicosity to obstruction, with venous repletion, through excessive muscular exercise.

2. *Boyer*, to obstruction, the gravitation of the blood, and weakness in the vessels.

3. *Briquet*, repletion from excessive muscular exercise, catamenial derangement, want of that support to the superficial which is accorded to the deep veins, diseases of thoracic organs, and mechanical obstruction.

4. *Begin*, adult age, muscularity, long standing, mechanical pressure on abdominal and pelvic trunks, disease of thoracic organs.

5. *Colles* and *Rokitanski*, inflammation of the vein coats, skin affections and injuries.

6. *Boileu* and *Andral*, excess of capillary action in the part from which the diseased veins arise.

7. *Bouillaud* and *Nelaton*, impaction of the bowels, principally the colon, and its pressure on the trunk veins.

8. *Herapath*, constriction at the saphenous openings.

9. *Chelius*, obstruction by gravid uterus, engorged bowels, or, gravitation of the blood in consequence of the erect posture, or without hindrance, weakness of the vein walls.

10. *Pigeaux*, to direct venous and arterial anastomosis.

11. *Verneuil*, in the first place, to dilatation of the deep veins; then to obstruction at the muscular and aponeurotic foramina, valvular insufficiency and injuries.

12. *Brodie*, *Skey*, and others, weakness of the vein coats in common with general laxity of fibre.

13. *Chapman*, prolonged obstruction from any cause; prolonged muscular action; weight of the blood; debility in the vein coats, and spontaneous or excited chronic phlebitis.

14. *Nunn*, pressure of the blood from within; first, from the force with which the blood is propelled onwards; secondly, from its pressure on the walls :—the first operating when obstruction exists at the trunk veins; the second when the walls of the vein are weak, or have been overtasked by too long continued muscular exertion.

15. *Sistach*, first, obstruction at the muscular and aponeurotic foramina; secondly, fatigue from long standing, especially combined with exposure of the limb to moisture or heat.

16. *Callender*, heritage, obstruction by thoracic disease, long standing, and gravitation of the blood; these causes operating with increased force if the veins are enfeebled by constitutional or other causes.

Add to these, causes connected with the constitution—age, sex, temperament, fevers, rheumatic and gouty inflammation of the veins, according to Warren, Mackenzie,* and Paget,†—and but few agencies that could possibly affect these vessels, so as to have the effect of developing varicose or scarcely any other disease, have been excluded from the category.

The return of the blood from the transpelvic limb, after completing its arterial and capillary courses, is, I need hardly say, accomplished by a somewhat complex machinery and a combination of forces. The veins may be divided into three sets, viz., the *superaponeurotic*, or saphenous ; the deep, or *subaponeurotic ;* and the *intercommunicating*, or those which bring the two former into direct connection with each other by perforating the intervening aponeurosis.

The propositions which I have to make with respect to these vessels, and which I present here as an apology for venturing to weary you, as I shall infallibly do, with a lesson from the "horn-book" of our science, are,—

1st. That there is, in the arrangement of the venous system of the lower limbs, as of other parts, so far as the trunk veins and their principal branches are concerned, a general and constant uniformity of plan or type ; and that the deviations from it, in any given instances, are not so considerable as to constitute in any one an exception to it.

2nd. That those *branches* which so far deviate from, as to be altogether irreconcilable with, such typical plan are supplemental and auxiliary ; and will have opened up, in all probability,

* Lettsomian Lectures.
† St. Bartholomew's Trans., vol. ii.

from inconsiderable vessels in obedience to some especial exigency of the circulation; and,—

3rd. That the veins which become varicose are not generally of the subordinate character just alluded to, but ordinary and regular branches which have important offices to fulfil, and become diseased through being overtaxed by excess of functional requirements.

These positions are derived from a careful and comparative study of the saphenous veins and their branches in their normal, as also in their diseased state; and they lead irresistibly to the conviction that in most instances the accurate tracing of a varicose vein from its known relations to trunk veins will serve to indicate the seat at least of the disturbance in which its disease originated.

I will now attempt to describe these trunk veins, and such of their branches more especially as are, from their character as intercommunicating vessels, most prone to become varicose.

1. The *internal* or great saphenous vein (Pl. I., fig. 1, I S) may be said to commence on the upper and outer border of the last phalanx of the great toe, where it sends down branches of communication with the deep veins. It passes along the edge of the foot, crosses the malleolus and inner surface of the tibia obliquely to its inner edge, which it reaches at the junction of the lower with the middle third of the bone. It then sweeps by a gentle curve along the inner belly of the gastrocnemeius to the posterior edge of the inner condyle of the femur, just over the tendon of the gracilis; and from this point to the saphenous opening, where it passes, by a sharp curve beneath the falciform edge of the fascia lata, into the femoral vein; or rather, into a seeming cloaca which receives

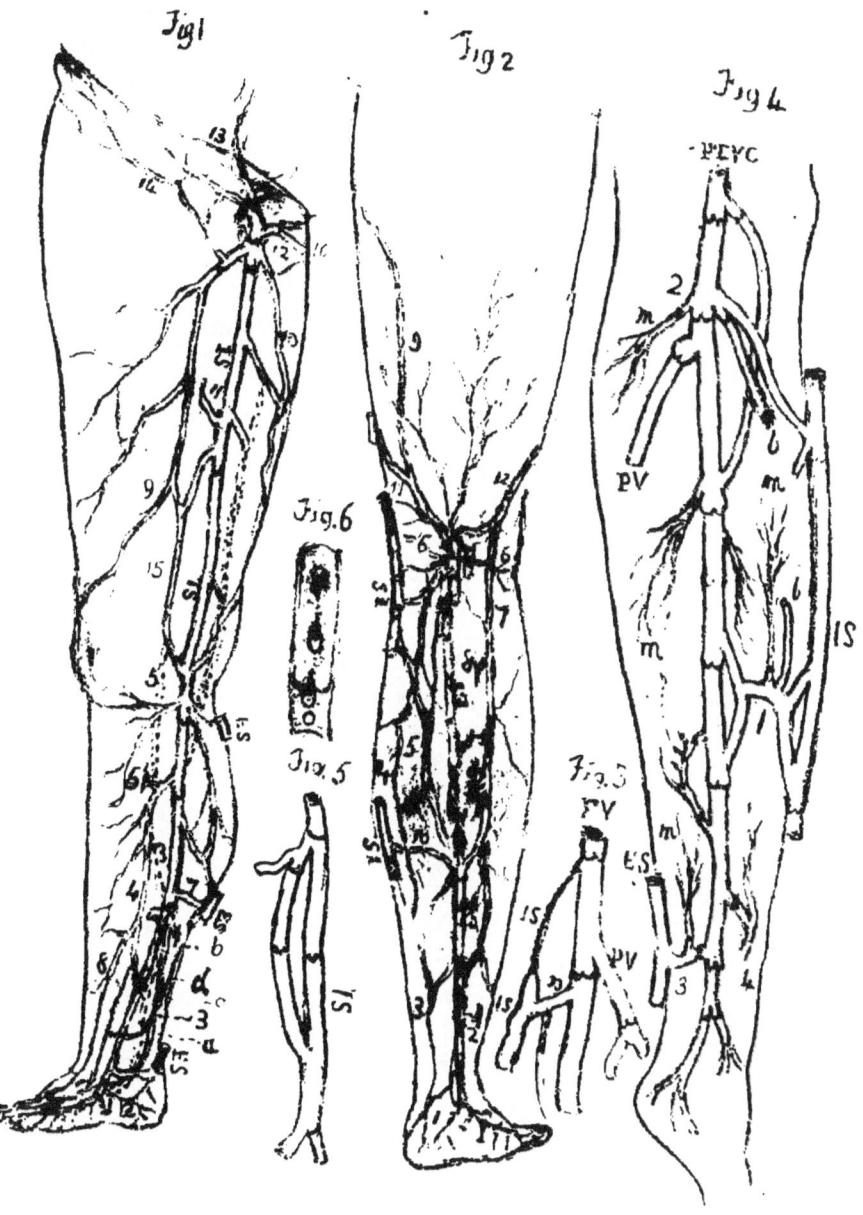

both veins as well as other important branches. In this latter part of its course the vein is closely surrounded and firmly gripped by the mesh of dense fibres which constitute the cribriform fascia. This is its most constant course; but it occasionally passes below and behind the malleolus (a, Pl. I.,) instead of crossing it. It very rarely deviates from the one or the other of these courses.

At first it lies close upon the periosteum of the several bony textures which it crosses; and subsequently upon the deep fascia, or rather, amongst its upper layers. In its course along the side of the knee joint the vein is more firmly bound down to the fascia than elsewhere by firm prolongations, which serve the purpose of fixing the vessel in its situation during the movements of the joint. It is often double in some part of its course; most commonly in the thigh, and just before its termination. It can scarcely ever be seen, and with difficulty felt, in either the living or dead subject, excepting where it crosses the tibia, and by the knee.

2. The *external* or lesser saphena commences in a similar manner, on the last phalanx of the little toe (Pl. I., fig. 2). It passes behind and below the external malleolus in company with the nerve which coasts it on its outer side, crosses the outer edge of the tendo Achillis obliquely, and reaches the middle of the leg just below the calf. Hence, we are told, it pursues its course above the tense fascial aponeurosis which covers the gastrocnemeius muscle, to the popliteal space, where there is an aperture in the fascia by which it finds its way to the popliteal vein. In all anatomical drawings, not excepting the most recent,* this said aperture is figured,

* See Bonany and Beau, Gray, and others.

and its dilatation is said to have afforded temporary relief to veins alleged to have become varicose through constriction by its edge.

I have made many dissections of these parts, but have not succeeded in finding such an arrangement in any one.

Arrived opposite the junction of the gastrocnemeius muscle with its tendon, this vein invariably penetrates the fascia, either by gradually insinuating itself amongst its fibres, or by the provision of a distinct foramen ; and from this point to its termination its course is unquestionably *sub*aponeurotic, and occasionally even *intra*-muscular (ES 11). In an instance in which the vein took this course, it was small in its intra-muscular portion, but its size was compensated for by two large veins which passed from it, below the point of its penetrating the muscle, to the internal saphena in the thigh (Pl. II., fig. 6).

The foramen closely encircles the vein, and often presents against its upper wall a very decided curvilinear edge; whilst the fascia itself is tough and unyielding ; so that any tendency on the part of the vein to yield at this point for the accommodation of a preternaturally swollen current is thereby inevitably counteracted. This foramen and its situation help most materially to account for the occurrence of crural varicosity in a large number of cases.

In their course along the borders of the foot, the two saphenæ inoculate most *freely* with the plantar and dorsal veins and with each other; and thus form a system by which the veins at the *extremity* of the leg have great freedom of collateral communication with the trunk vein at the middle and termination of its course.

Tracing the *internal* saphena upwards, its more constant branches (those which are chiefly engaged in providing channels of communication between the remote parts of the same trunk vein, or between the superficial and deep systems) are,—

1st. A branch (2) below the malleolus to the external saphena, which forms a curve, from the convexity of which a branch (1) passes to the plantar vein.

2nd. As it passes over the malleolus, a loop vein (3), which, after receiving tributaries from the heel and sole, and giving branches to the plantar vein, runs up behind the inner edge of the tibia, either to rejoin the trunk at the middle of the leg, or to pass on to the next branch, the *long dorsal* vein. This loop vein sends a branch (*a*) across to the internal saphena, or to the long dorsal above the malleolus; a branch (*b*) to the posterior tibial, about the junction of the soleus muscle with its tendon— the inferior *perforating* vein on the inner aspect of the limb which often receives the soleal vein; and (*c*) an important branch to the external saphena below its aponeurotic opening.

3rd. The *long dorsal vein** (4) completes the inner cornu of the dorsal arch of the foot, and after communicating on its dorsum by large branches and freely with the anterior tibial, the saphena, and plantar veins, ascends the leg on the outer side of the spine of the tibia, crosses that bone about its middle, and joins the saphena behind its inner edge. It does not, however, always terminate here; but sometimes prolongs its course either just behind or in front of, and parallel with, its trunk, to

---

* This branch is very important. It was called by the older anatomists the "vena minor saphena," but it has no claim to be so called, for it has none of the distinctive features of the saphenous trunks.

unite with it in the middle or higher part of the thigh, or with branch (5).

After crossing the tibia in the middle of the leg, and about an inch from its junction with the saphena, the long dorsal vein receives the upper *perforating* branch on the inner aspect of the leg (6);—a branch which connects the internal saphena with a posterior tibial vena comes, either by traversing the soleus muscle close to its attachment to the tibia, by a very zigzag course between its muscular and aponeurotic fibres, or by passing beneath its insertion. It is accompanied by an artery and a nerve. Occasionally this vein deviates from this course, and takes the last part of its course behind the saphena (see dotted vessel), in which case it often receives the loop vein (3).

4th. The *short dorsal* (8), which completes the outer cornu of the dorsal arch, and, after freely communicating with the external saphenous and plantar veins, passes up in front of the fibula, and ultimately either unites with the former vein, about midway up the leg, or takes an independent course to its trunk. This vein also sends a branch to the posterior tibial, just before its termination,—the middle *perforating*.

5th. A considerable branch, which arises deep in the tissues on the outer side of and immediately below the knee joint, where it communicates with the deep articular veins (6, 6, fig. 2), and, crossing the leg, unites with the saphena on its inner side. It receives branches, which cross the tibia obliquely from the outer side of the leg, where they anastomose with branches of the external saphena.

6th. A branch (5) disengages itself either from the long dorsal vein, the loop vein (3), or from the saphena itself, just above the knee joint, and runs upwards, taking nearly the same course as that vessel, but on a higher plane, to join it just before its termination.

In the thigh the saphena usually receives three anterior and as many posterior branches :—

1st. A branch (12, Pl. I., fig. 2) which disengages itself from the popliteal, crosses the outer side of the knee joint, and after continuing its course obliquely above the patella, joins the saphena in the middle or upper part of the thigh. This branch is joined by an important tributary (7, fig. 2), which, after leaving the saphena below the calf, sweeps along the outer belly of the gastrocnemeius, and in this part of its course sends down two branches to the posterior tibial venæ comites (8, 8):—the superior and middle *perforating* veins on the outer side of the leg.

2nd. A vein (9, fig. 2) taking a somewhat analogous course from the inner side of the popliteal space, where it communicates directly or indirectly with the popliteal vein, and after winding along the inner side of the knee joint and aspect of the thigh, unites with the saphena either in the middle of its course or at its termination,—in which case it occasionally appears as a continuation of the external saphena (10, fig. 1),—or at both points. In one case the vein assumed this character, and gave, at the situation where the external saphena usually joins the popliteal, a very trifling branch only to this vein; whilst in another like case the connecting branch was very large.

3rd. A branch (11, fig. 1) from the saphena, about the middle of the thigh, which passes obliquely upwards and inwards to join the femoral vein opposite the mouth of the profunda. This is an important inosculation, for it is the channel by which the blood of the saphena finds an outlet when its orifice is obstructed. In one dissection I found such an arrangement of these vessels (see 10, Pl. I., fig. 3). The saphena (I S) was large up to the point of giving off this branch—its channel was in effect obliterated—and *very small* beyond. This cross branch receives a vein which traverses the semimembranosus muscle to join the external saphena in the popliteal space. It is often tortuous; and in a case to be mentioned hereafter a diffused varix formed in consequence of its rupture in the body of that muscle. These branches inosculate with each other and often form, with other accessory branches, a somewhat complicated network on the front and inner side of the thigh, in connection with the trunk vein in the latter part of its course.

4th. The external circumflex (14) superficial pubic (12) and epigastric (13) branches inosculate respectively with the internal pudic, obturator, vesical, and uterine veins; with the deep epigastric; and with the ileo-lumbar veins. Through these and other intermediate veins—as the azygoid, internal mammary, and axillary—the femoral vein may be brought into direct connection with the superior vena cava.

The internal saphena has usually but seven valves;—one over the tarsal bone of the great toe, below the branches which it gives to the plantar veins; another by the ankle; a third above the malleolus; a fourth within a short distance below the knee joint; a fifth about midway up the thigh; and the remaining two in the last portion of its course, at the distance of one, two, or three inches respectively from its confluence with the femoral. The mouth of the femoral at the confluence of the saphena is valved, but the mouths of the saphena's tributaries, above its last valve, are usually not so guarded. The second valve of the femoral after the embouchure of the saphena is from an inch to an inch and a half above it; hence in case of any obstruction causing regurgitation of the current in the femoral vein, the blood would be, as it were, shunted into the last segment of the saphena; and thus, after a time, what is usually termed a femoral, but what is really a saphenous varix, is formed. This is, as far as I have been able to discover, the more general distribution of these valves; but it is by no means uniformly the same, nor are the valves always placed behind the mouths of large branches.

The branches of the *external* saphena (Pl. I., fig. 2) to which it is necessary for me to direct your attention are—

1. Several small veins (1) which inosculate with the plantar, dorsal, and posterior tibial veins on the outer side of the ankle joint.

2. Above the ankle, a branch (2) from the peroneal or posterior tibial vein,—the inferior *perforating* branch.

3. A branch (3) from below its foramen downwards and inwards to unite with branch (2) of the internal saphena. This branch gives usually a connecting branch to its trunk in the popliteal space,—a duplicate of that portion of the main vein.

4. A cross branch (7, fig. 1; 10, fig. 2) from the same portion of the vein to the internal saphena, or to its long dorsal branch.

5. An important branch (5), often as large and as thick as the trunk vein, which takes its rise also just below the foramen, and takes a course upwards and inwards to the internal saphena behind the knee joint. This branch is often entirely subaponeurotic.

6. A still more important branch, which ascends from the last portion of the vein to the internal saphena, a short distance above the knee joint.

These veins anastomose with each other with more or less freedom in different limbs by branches which, having no regularity of distribution, may be regarded as supplemental.

The perforating veins are, for the most part, difficult to be traced, unless they have been developed by conditions which have given them more than ordinary importance in the circulation.

All these branches are prone to varicosity; but, as we shall see, there are certain of them which have a special tendency to become varicose.

The external saphena has an indefinite number of valves —from three to nine as in this vessel (fig. 2). In this instance there was a valve below the orifice of each branch that was visible to the eye.

This, as well as the internal saphena, but more frequently the former, is occasionally a double vein in some part of its course, with lateral intercommunication.

The *branches* of the saphena, with exceptions to which I will refer, are destitute of valves. I make this statement after a careful examination of these vessels in a large number of bodies; for it is generally supposed that the contrary is the fact, so that varicosity is rarely brought under clinical notice without a reference to valvular implication, and especially to the so-called reliquiæ of degenerate valves.

Valves are, however, occasionally met with in the larger branches, but they appear *functionally* to appertain to the trunk vein. For instance, I have found them thus disposed :—

In one case the saphena terminated in the femoral by two veins, running parallel and close to each other. In one there was a single pair of valves; in the other, two pairs. It is obvious that one of these veins was a duplicate of the other, the trunk vein; and that the valve in the secondary was a necessary adjunct to that in the primary vessel. In other bodies I have met with a valve in a loop vein, but only in case of there being a valve in the trunk vein in collateral correspondence with it, as in Pl. I., fig. 5. The reasons are obvious.

The perforating veins, of which I have described the more important, in their course to the deep veins pass through dense structures,—fascia and muscle; which present formidable barriers to the return of the blood after it has once passed them in the direction of the current. The foramina in the deep fascia are mere *slits*, of which the edges slightly overlap each other; whilst the passage through muscles at their attachments is often very crooked or zigzag, from the interruption which their aponeurotic elements offer to a direct course. Thus these vessels, as they thread their way from the sub- to the super-aponeurotic system, are virtually valved, and, as a consequence, the first traces of varicosity show themselves as varices at the sites of their fascial outlets. Here, too, the dilatation of these veins usually ends.

The courses of the deep veins, as they are comitial to the arteries, are too well known to need description in detail.

The veins of the foot inosculate so freely, and such are the means of escape for its blood, through either the superficial or deep system, that embarrassment to its venous circulation is hardly possible. It is not so, however, with regard to the superficial veins of the leg between the ankle and the calf. In this part of the limb the circulation is exposed to greater difficulty and risk of obstruction than in any other part of the limb, if not in any other part of the body.

The deep venous system of the leg is wonderfully arranged so that the muscles shall suffer no embarrassment to their functions from congestion. The addition of a duplicate vein so far as the popliteal, and occasionally, as I have seen, as far as the profunda; the abundant means of communication between these comitial veins; their connection with the saphenous system; and their valves, are amongst the means to that end. I must ask your attention to a few facts in detail.

If a vena comes—say a posterior tibial—be slit up, two kinds of orifices or mouths will be seen (Pl. I., fig. 6):—1st, those by which one vein opens into the other; and 2nd, others which are the mouths of the muscular and perforating branches. The former are club-shaped, the latter round.

The openings of the first class communicate between the two veins, either directly; or, by short channels as though formed by a duplicature of the vein wall, which, in the form of a partition, shuts off a small portion of the cylinder. In case the channel is direct, the *apex* of the club-shaped opening is pointed in the direction of the stream; but when the communication is by an intermediate passage the upper orifice has its apex in that direction, whilst the apex of the lower is reversed, as is seen in the diagram referred to.

The *crescentic* portion of each orifice is endowed with a filmy, cusp-like fold, which admits of slight movement. By this arrangement the stream, if diverted from one vena comes into the other, is constrained to take an onward course. It cannot regurgitate. These openings are sometimes situated between the valve cusps.

The arrangements by which the venæ comites and the saphenæ intercommunicate are also obviously intended to subserve the same purpose, and are of sufficient interest to deserve notice.

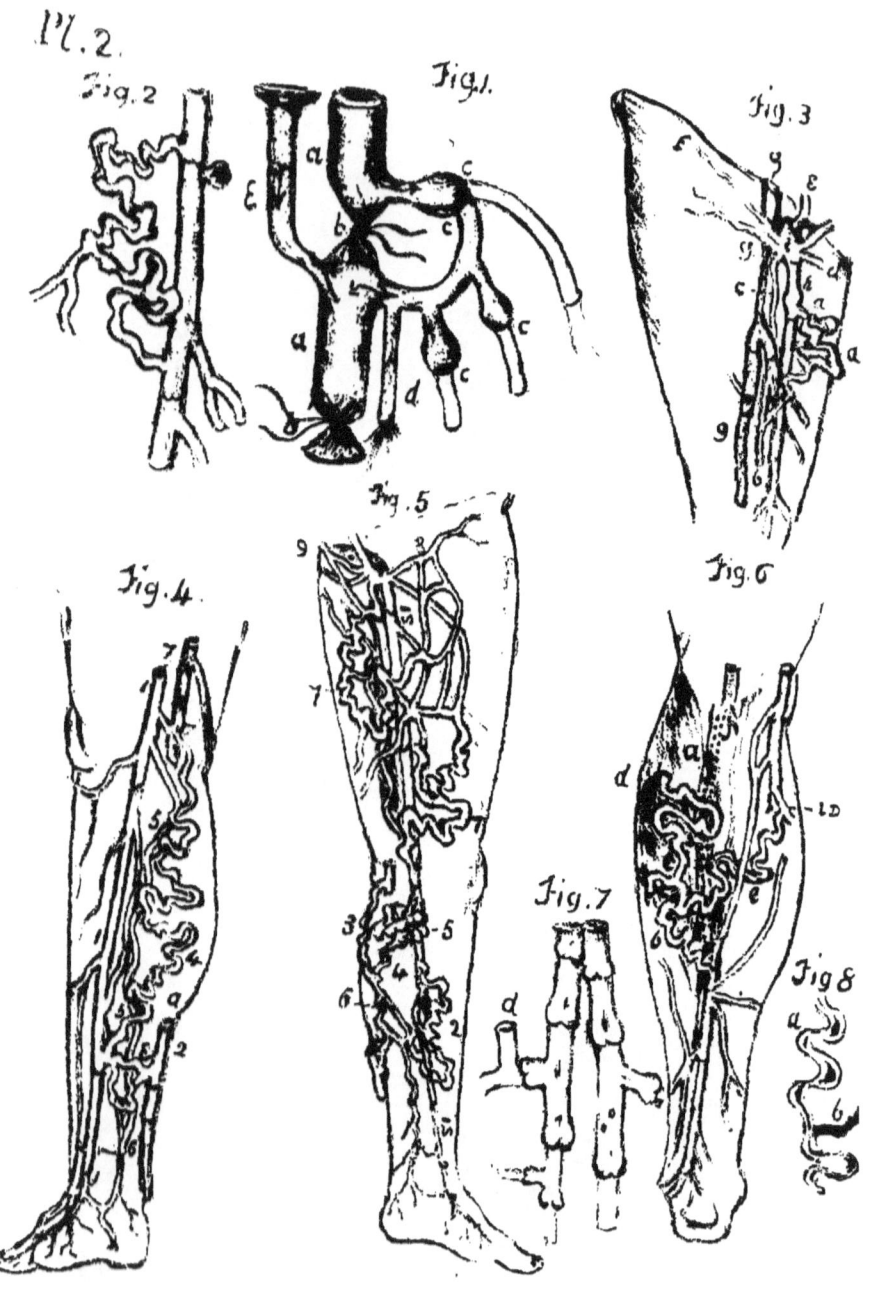

They are illustrated in Pl. I., fig. 4. At 1, a transverse vein is seen to connect a vena comes and a branch of the internal saphena by subdividing at either extremity into two branches of its own size. At the point where the subdivision takes place on the saphenous side, a branch from the bone (*b*) and a muscular branch (*m*) enter concurrently. At 2, the branches from the bone and muscle respectively enter the vena comes opposite the mouth of the transverse branch by which it communicates with the saphena. At 3, a muscular branch enters the connecting branch midway between the two. Such arrangements are frequent, but not constant, as may be seen in the same illustration—taken from a dissection;—and are obviously designed to facilitate the escape of the blood from the muscular and osseous veins, and to insure its diversion from the deep to the superficial system in case of need.

Again, the branches of the muscular veins are abundantly supplied with valves, and these valves are so arranged as to insure for the currents a centripetal course. In the illustration (Pl. II., fig. 1), *a a* represents a portion of the femoral vein from the human subject, ligatured at *b; c c c c* branches, valved below their bulbous portions; *e* a glass tube, inserted into the lower segment of the vein. Mercury poured into this tube filled the lower, and immediately rose into the higher segment of the vein; its escape being prevented by the valves in the branches. From branch *d*, however, the mercury escaped. It is probable that either there was a valve at a point below where this branch was cut, or that it was a communicating vein.

This arrangement found throughout the veins of the muscular system forms a safeguard against intra-muscular dilatation, and explains the fact, which I shall hereafter more fully advert to, that congestion of the *muscular* veins is a condition neither easily nor readily procured. It shows, moreover, that relief from impending congestion of these veins is to be derived from the diversion of the current into the saphenous system; and offers an explanation, too, of the fact that œdema of the leg is confined almost entirely to the superaponeurotic textures. Still, the intramuscular veins are sometimes found dilated; and I cannot but think that with their dilatation ordinary cramp is directly associated.

In continuing the anatomy of the veins I may just remark that the popliteal receives many large muscular branches, and *that at a point* where it has considerable communications with the saphena. It also communicates with the pelvic veins.

The walls of the adductor opening, through which the femoral vein and artery pass, are not only firm and unyielding, but closely environ the vessels, and present a firm crescentic edge against them posteriorly.

At its termination, again, the femoral vein is firmly encompassed by the cribriform fascia and other structures within the canal. The fibres of the former are tense and unyielding, and their tension on the veins is increased by the act of extending the thigh on the pelvis as in walking. These two apertures appear to act as diaphragms. They regulate the size of the stream, when swollen beyond the normal capacity of the vessel; and *that* to the embarrassment of the veins below, but for the provision made by the intercommunicating veins through which the surplus blood may be diverted into the saphenæ. Inosculation of the branches of the muscular with other veins does not materially contribute to the relief of obstructed venous circulation in the subaponeurotic system.

But the saphenous trunks are, as we have seen, intercepted by like diaphragms, and these have obviously an important bearing upon the fixed and, at times, inadequate outlet for the blood provided by the femoral vein.

In case of an excess of blood in the deep or muscular system, anatomy shows, therefore, that provision is made for its discharge into the saphenous; but, as we shall see, its veins have no means, direct or collateral, of relieving themselves of the surplus; consequently, they act in the capacity of *reservoirs*, with languid currents, until they are relieved by a general subsidence of the stream.

The branches of the saphenæ, in contradistinction to the trunks, are interwoven with the dermis in such a manner that alternate segments are almost entirely encompassed by its fibres, whilst those which intervene are free with the exception of their attached upper portions. This appears to arise from the fact that the vein skirts the opposing confines of the dermoid and adipose layers, and, in consequence of this arrangement, passes through the areolar walls of the fat cells as well as through the cells themselves, where they are in direct contact with the fat. Hence, in its course along the side of the knee joint—or elsewhere where the fatty layer is comparatively scant and often deficient,—the vein is bound by an infinite series of small areolar plaques to the fascia as well as to the skin, and is thereby fixed during its movements. It also follows that when a vein dilates, those segments first yield which occupy the fat cells, and that in the unattached portions of the cylinder only. It is easy to dissever a varicose vein from its adipose, but not from its areolar or dermoid connections.

The strength of the vein coats—*i. e.*, its power of resisting tension varies in different veins.

A portion of healthy saphena will sustain a weight of little more than five pounds, whereas a portion of one of its largest branches of equal length is only equal to the sustentation of a weight of about three pounds four ounces. Its resistance to columnar pressure, in some experiments which I made, was found to be equal to that of a column, differing in different veins, of from twelve to eighteen inches of mercury; whilst a portion of the large saphenous branches bore a column of from nine to twelve inches only, and portions of the posterior tibial vein yielded to a column of from eight to twelve inches. In making these last experiments I was struck with the fact that when the veins yielded, they furnished exact resemblances of those various forms of dilatation which they respectively exhibit in connection with varicose disorder.

But to this advantage, in point of physical strength, which the saphenous trunks possess, in comparison with their branches, must be added that also which they derive from their valves. And the same comparison might be made between these branches and the veins of the subaponeurotic system, through which valves are abundantly distributed. Hence, physically and functionally the superaponeurotic *branches* of the saphenæ form by far the *weakest* department of the venous system.

I must now refer briefly to the powers by which the venous circulation in the lower limb is carried on.

Modern physiology is much in favour of an almost purely physical explanation of this phenomenon. The experiments of Haller, Magendie, Ludwig, Sharpey, Volkman and others, seem to concur in showing that the force of the heart distributed by the arteries is sufficient to carry on the circulation, since the pressure of a 3½ inch column of mercury on the aortic column above the renal arteries will return the blood after its entire circulation in a full stream to the cava above the diaphragm; whilst by increasing the height of the column to five inches, the blood spurts from the vein in a full jet. In the human body two or three days after death, the circulation can be, as I have experimentally satisfied myself, carried on by the substitution of a small two-valved pump for the heart, and by a force equivalent to that of the pressure of five and a quarter pounds, even when the impetus is communicated to the current in the first portion of the femoral artery.

But this is perhaps a too cardiac view of the circulation; for other experiments by Bernard,* Williams, and especially those by Dr. Draper,† go to show that the impetus of the blood is liable to modification in its passage through the capillary system; and assuredly, as Magendie alleged, it is augmented by the elasticity of the vein coats themselves.

But besides these there are other agencies which are capable of modifying the force of the venous current. What are we to say of the valves?

---

* Bernard showed by his experiments of galvanizing filaments of the sympathetic and the chorda tympani respectively, that the more difficulty the blood experiences in its course through the capillaries, the greater is the inequality between the arterial and venous tension; and *vice versa*. Thus, with relaxation of these vessels venous tension is increased; and it is not improbable that to this cause the phenomenon of venous pulsation, as first detected by Mr. King by means of the beautiful contrivance of a thread of sealing-wax—the precursor of our modern sphygmograph, is due. These experiments give some colour, too, to the views of Pigeaux in reference to the *cause* of varicosity, to which allusion will hereafter be made; but it can hardly be surmised that an amount of force can be generated in the capillaries sufficient of itself to account for varicose dilatation.

† See Dr. Draper's "Physiology" on the subject of Capillary force.

The primary and main function of the valve, in the *normal state of the circulation*, I have already said to be that of guarding the current against divergence from its normal course; and by no means, as alleged, to assist the walls of the vessel in sustaining the weight of a continuous column of blood. So long as the stream takes its normal or centripetal course, the blood fills the vein, and the cusps of the valves are fixed against its sides. The vein, although valved, forms an uninterrupted channel; the cusps can only interfere to break the continuity of the column by pressure on their upper faces; and this can only happen through oscillation of the stream in consequence of some interruption to its progress,—through, in fact, an *abnormal* state of the circulation. In such a contingency the first valve below the point of oscillation would close; and the stream below would thereby be diverted into a collateral branch. The circulation would proceed with one valve closed through the pressure of a superincumbent column of blood in a condition of comparative stasis, unless the cause which called the first valve into obstructive operation should extend beyond the circulating area over which it presided, when a second or other valves in succession would be called into action. Hence, the valve cusps are called into action only in the event of regurgitation of the current. In this case they have the effect of preventing further reflux, and insuring, for the divergent current below, freedom from the embarrassment to which the column above is subjected. Under like conditions the unvalved veins have no such resources to fall back upon. Hence the saphenous trunks are to a certain extent provided with powers of resisting the effects of exalted blood tension which their branches do not possess; and the latter will yield to derangements of the circulation which upon the former would fall comparatively harmless. But there is a portion of the vein upon which valvular action has negatively a weakening influence: viz., *that* above the closed cusps. It is here that the strain of the superincumbent column and of the force which resists its progress is felt; and at this point under such conditions dilatation is constantly met with; whilst below the valve, the vein receives concurrently an accession of strength.

It is unnecessary to direct your attention to the auxiliary influences of the thoracic movements, of the muscularity of the cava, and other analogous agencies, in promoting the return of the blood from the lower limbs. The former I shall have occasion to notice hereafter.

I have said enough, I trust, to show—and this is my object—that whatever be the special disadvantages to which the venous current in the lower limb is subjected, the machinery and forces employed are, in their normal state, not only amply sufficient for its maintenance, but leave a margin for possible embarrassment from the operation of any extraordinary opposing forces that can be evolved exclusively *by the organs of circulation.* This margin is thus expressed by the calculations of Volkman, Ludwig, and Mogt, as far as merely *blood tension* is concerned. According to the former, the blood in the veins retains 1-300th part of the tension impressed upon it by the heart's systole, independently of what it gains *in transitu;* whilst the latter affirm that its tension in the carotid of a dog, compared with that in the jugular vein, is as 5·09 to 0·58.

This is altogether insufficient to effect dilatation of the vein walls; for, as we have seen, in order to this the force evolved must be equivalent to that exerted by a column of mercury averaging from ten to eighteen inches in height.

It follows, therefore—and this is the practical point,—that neither systolic, arterial, capillary action, nor columnar pressure can singly, or even combined, evolve sufficient force wherewith to overcome the resistance that the vein walls in their healthy state can oppose to them; so that we must look for such a force beyond the domains of the system in which it displays itself.

And I must at once premise that a force capable of determining such a result can only be obtained from the muscular system. There is no other source to which we can look for it; and the reciprocal arrangements and relations which exist between this and the venous system, explain the mode in which any force, generated in the one, can be communicated to the other.

What, then, is the relation of the saphenous system to the muscular, and both to the deep venous system? Mr. Nunn remarks that the saphenous veins "offer to the returning blood an unobstructed passage during the time the venæ comites are undergoing compression by the contraction of the muscles of the limb; and that they thus act as a safeguard against the contingency of congestion of the active organs of locomotion. One may compare these veins to the adjustment of the chronometer, both having for their object a uniformity of performance under varying conditions."*

In order to determine this point I made the following experiments, with the able help of my friends Dr. Richardson and Mr. W. Smith. A stream of defibrinized blood, at the temperature of

* Nunn's Essays.

98° Fah., was pumped into the femoral artery of a fine muscular subject, recently dead, when it speedily passed into the femoral vein. The action of the pump was continued, but with regular intermissions, so as to resemble that of the heart. The arteries pulsated; the limb gradually became rounded and shapely, the veins swelled; the skin assumed a flesh colour; blood flowed from the vessels and skin when pricked,—from an artery, in a jet. In short, the dead limb became a striking mockery of its living self. In this experiment the skin capillaries, in certain regions, became injected prior to the repletion of the subcutaneous veins.

I repeated the experiment after tying the internal saphena at its termination, and with the following results:—First the superficial veins began to swell, in order from the foot upwards; then some small tegumentary veins in front of the inner malleolus; and subsequently the long dorsal vein. A wound in the opposite limb bled freely. On examining the limb, the saphenous system was found generally to have been injected as well as the deep trunk veins; but the intra-muscular veins and muscular tissue showed no sign of injection.

On another limb the same experiment was tried, but the process of injection was prolonged beyond that in the former experiment. The results were as follows :—

The saphenous veins and their tributaries to the dorsal veins of the foot were first filled; then, in order, the capillaries of the skin of the toes, of the heel, and of a circular portion of skin above the patella. The muscles of the calf were *slightly* tinged with the injection.

The opposite limb was then injected in like manner, *the femoral vein being tied on the distal side of the saphenous orifice.* The results differed, and occurred as follows :—

The saphenous veins and their tributaries to the dorsal veins of the foot were first filled; then, in order, the capillaries of the skin of the toes, of the heel, and of a circular portion above the patella. The femoral vein became immediately filled to the ligature, and the wound bled; the *skin* around the patella became turgid, the turgidity commencing at points and extending itself peripherally, until the whole patch became uniformly filled; then, in succession, that of the toes, heel, and sole of the foot; next, the branches of the saphenæ, and subsequently a broad patch of skin along the course of the inner saphena, from the toe to the knee; the skin along the outer edge of the foot, with tributaries running in the direction of the saphena; the two upper perforating veins, and, finally, some patches of skin along the inner part and front of the thigh.

In a fourth experiment, blood, prepared as before, was *very rapidly* injected into the femoral artery of a man who had been dead thirty-six hours. The blood soon made its appearance at the saphenous and femoral orifices, and then simultaneously in the sub-tegumentary veins and capillaries of the skin around the patella. The skin along the front of the thigh next became rapidly tinged, the turgidity beginning, as before, at distinct points, separated by considerable spaces, and radiating irregularly in all directions until the peripheral boundaries of each distinct patch met, and the whole was uniformly injected. The injection then extended rapidly to the buttock, on the one hand, and to the skin overlying the internal saphena on the other, until it reached the tributaries along the inner edge, and, finally, the skin overlying the external saphena and its tributaries on the outer edges of the foot. The skin of the toes and sole of the foot became injected just in advance of that overlying the trunk of the saphena. On examining the tissues of the limb, I found that the whole of the saphenous system had become intensely injected, and likewise the femoral, the iliac veins, and the vena cava to the heart; whilst the anterior and posterior tibial veins and the peroneal were comparatively empty, and the muscular tissue altogether uninjected.

I repeated these experiments again and again in the presence of Mr. Jordan of Manchester, Dr. Randall, Dr. Southey and others, and with the same results, excepting, perhaps, some trifling and unimportant variations in the order in which the different parts became injected.

These observations, if they yield anything, which I cannot doubt, seem to show that obstruction of the femoral and internal saphenous veins respectively is followed by different categorical consequences to the circulation in the limb.

Obstruction of the femoral vein is followed in succession, according to the order in which the results are given, by,—

1st. Circumpatellary cutaneous injection;

2nd. Saphenous repletion;

3rd. Repletion of the intercommunicating veins;

4th. Capillary cutaneous injection along the course of the superaponeurotic veins,—first of the thigh, then of the internal and external saphena in the leg; whilst—

Obstruction of the internal saphena is followed by,—

1st. Repletion of its whole system;

2nd. Repletion of the veins passing between the saphenous trunks below the calf;

3rd. Injection of circumpatellary skin; and,—

4th. Of the deep venous trunks in part.

In all the experiments the saphenous system was very speedily filled, and in intensity according to the rapidity of the current. When the femoral vein was deligated, the cutaneous veins of the thigh around the patella and along the leg were the *first* to indicate the obstruction; whereas with deligation of the internal saphena these particular sets of venous capillaries were, in order, the last to be injected, and followed the injection of the saphenous veins to their capillary radicles.

Hence it would appear that tegumentary (*veno-capillary*) and sub-tegumentary (*superaponeurotic venous*) injection become severally notable and discriminating signs. The one—the tegumentary—denotes obstruction of the femoral vein; the other, obstruction of the saphenous.

I had scarcely concluded these experiments when I accidentally fell upon a monograph by Dr. Sucquet, entitled "D'une Circulation Dérivative" (1862).

Dr. Sucquet had been struck by the fact that the saphenous veins were not constantly filled, but could be filled by exercising the muscles of the limb or exposing the limb to heat. In order to explain this circumstance Dr. Sucquet injected the femoral artery with a black alcoholic solution of resin, stopping the process as soon as he found the fluid had reached the dorsal veins of the foot.

The injection reached the plantar veins and the other parts of the venous system in the order in which they are given; viz., the capillaries and venous ramuscles which contribute to form the saphenous veins; some of the large branches of these veins below the knee; the network of vessels surrounding the patella; the portion of internal saphena adjoining the knee joint; the vein which passes from the external saphena below the calf to the internal behind the knee. These several parts were injected before the deep veins or the other veins of the leg, and by continuing the injection the blood reached the femoral before it reached the termination of the internal saphena.

Dr. Sucquet concludes "that there is in the lower limbs two distinct circulations: *one*, deep, nutritive, constant, and uniform; the other [which he terms "derivative"], unequal, intermittent, and which concentrates in the saphenous veins and some of the vessels of the knee;—a circulation of accommodation, or, according to Bernard, the mechanical and not the chemical circulation. Although the object of Dr. Sucquet's experiments did not quite tally with that which I had in view in those I made, yet the details in both so nearly correspond as to be reciprocally corroborative as to the results.

One of the most striking results obtained in the foregoing experiments was the almost entire immunity of the muscular system from penetration by the injection even in part to the comitial veins; the muscles of the calf showed *slight* indications of having been penetrated by the injection when the saphena was tied; and as the fact, if it could be accepted, would bear considerably upon the open question of intra-muscular dilatation, as the first step in the process to superficial varicosity, I endeavoured to decide it by an appeal to the living animal.

For this purpose I undertook the following experiments, in which I was most ably assisted by my friends Mr. Savory and Mr. W. Smith.

A dog was rendered insensible by chloroform, and a ligature put on the femoral vein, on the cardiac side of the orifice of the saphena.

The animal was killed on the seventh day.

There was enormous distention of both the internal and external saphenous systems, especially of the femoral portion of the former, to its minutest ramuscles. The femoral and popliteal veins were gorged from the entrance of the external saphena,* together with the branches of the profunda, but the deep veins of the leg were comparatively empty, and the intra-muscular veins entirely so.

The blood had been returned collaterally, in small part, by the branches inosculating between the pelvic and profunda veins, but mainly by the superficial pubic and epigastric, and the ileo-lumbar veins.

In a second animal the femoral vein was tied in the same situation. The dog was killed and examined on the tenth day, with nearly the same results.

The saphenous veins, with the femoral and popliteal, were enormously distended, *even to tortuosity of one large intercommunicating branch in the thigh.* Inosculating branches between the glutœal and profunda were also distended; but there was the same absence of repletion in the trunk veins of the leg, and the muscular tissue was everywhere as pale as that in any other part of the body. The limb was œdematous above the deep fascia.

In a third, the femoral vein was destroyed by caustic to the junction of the profunda. The dog was killed on the seventh day. Plate III., Fig. 1.

* The internal saphena joins the femoral considerably lower down the thigh in the dog than in the human being.

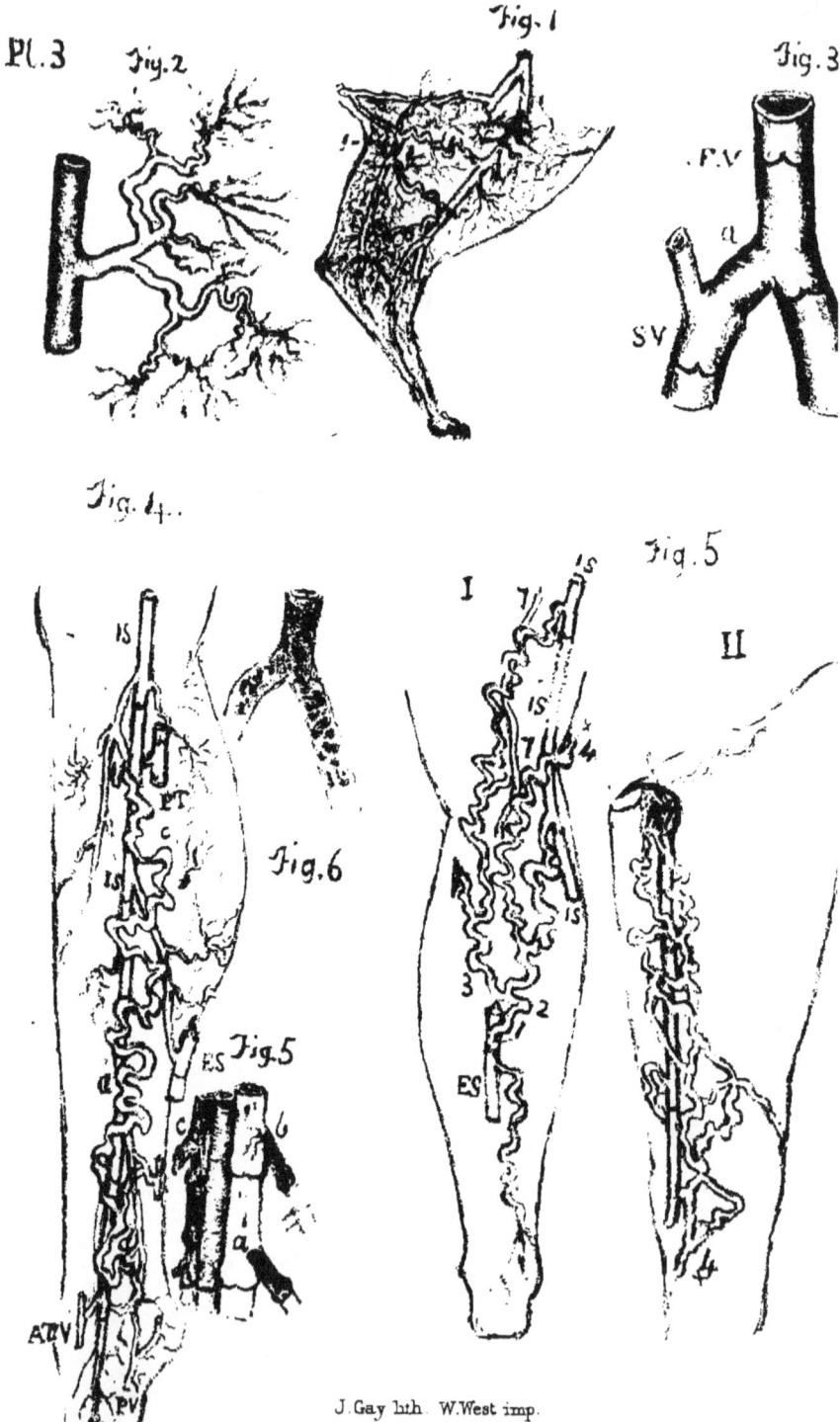

The capillary vessels on the outside of the thigh, and especially above and on either side of the patella, were enormously congested; and led to large and tortuous veins which ran to a large vessel which occupied nearly the same position as the saphena in the human subject, but taking a superficial course (1). This vein inosculated with the ileo-lumbar, and superficial epigastric veins, which were much enlarged and gorged. The vascularity ran down the inner side of the leg along the course of the distended saphena, and gradually disappeared, being no longer perceptible about midway. There was no muscular injection, with the exception of the muscles in the immediate vicinity of the destroyed portion of veins. These were very red, apparently from inflammation.

In a fourth experiment a ligature was placed on the vena cava of a large animal, which resulted in his death on the following day.

The vein to the ligature was distended with thick but not coagulated blood. There were marked efforts towards the establishment of collateral circulation. The superaponeurotic system of veins (the saphenous), to their remotest ramifications, were intensely congested. The intensity increased towards, and culminated on the foot, on each side of which it had gone so far as to have resulted in ecchymosis. The muscles were not at all congested, nor was there any remora of the deep veins, including the popliteal and femoral, below the valve which guards the femoral vein below the orifice of the saphenous. Above this valve the femoral was much distended. These experiments show that in the embarrassment to the circulation of the lower limb through obliteration of the cava, the saphenous system will become intensely injected in every part, without concurrent injection of the muscular system.

In order to render these results more reliable, I made one other experiment; repeating it several times with uniformity of result.

The whole of the thigh of a dog was encircled by ligature, so as completely to strangulate the vessels, with the exception of the femoral artery. As soon as the femoral vein showed signs of great distention the artery was tied, and the animal killed.

In the thigh the femoral and saphenous veins were equally and intensely congested; the former to its intermuscular branches as far as the nearest valves, the latter to its minutest ramifications. The venæ comites of the leg had but very little blood in them, and those of the foot still less. The saphenæ were enormously injected, even to the capillaries of the skin of the toes, heel, and sole of the foot; but perhaps rather to a less degree than the ramifications on the thigh. The muscular tissue and intra-muscular veins were throughout uninjected.*

These experiments show that obstruction of the femoral vein above the internal saphenous orifice is followed by,—

(*a*) Intense congestion of the whole saphenous system to its capillary derivatives;

(*b*) Extension of the congestion to the popliteal and femoral veins. This congestion diminishes, however, *in the deep trunk veins* gradually in the direction of the foot; so that, below the calf, the blood in these vessels is not only not in excess, but even short of its normal quantity; and

(*c*) Immunity of the *intra*-muscular veins and their ramuscles, even to the capillary system.

A collateral venous circulation, by means of inosculation

* To avoid misapprehension I may state here that I have used the terms *inter* and *intra* in relation to the veins, to distinguish between those portions of the veins respectively that are *between* and *within* the muscular structures.

between the veins of the pelvis and thigh, does not appear capable of materially relieving the veins of the lower limb from the consequences of over-repletion.

Now these *results* tally in almost every respect, as we shall hereafter see, with the state of the veins in cases of varicosity. Their preliminary conditions do not, however, appear *at first sight* to be analogous, and if so, no inference of any value can be obtained from them; but it is not so, for the actual ligature in the experiments given is capable of being simulated, so far as its effects are concerned, by augmentation of the current in the saphenous veins.

It is well known that the quantity of fluid which at any time traverses different sections of a tube is always the same; so that when a tube diminishes the velocity of the current increases, and with it, the tension of the contained fluid, and *vice versa*. If we examine the circulation in the saphenous system, as it has been described by the light which this law sheds upon it, we shall not fail to discover that there are possible conditions under which the current might be *passively* obstructed at the saphenous or femoral orifices—and, indeed, at any other foramina through which these veins pass,—and as effectually as though the obstruction had been caused by a ligature. And for these reasons. The capacity of these orifices, on the one hand, is fixed and invariable, and so practically is the calibre of the saphenous veins; whereas the sum of the capacity of their *branches*, on the other, *far* exceeds that of their trunks respectively. It follows that, *with the ordinary current*, the velocity of the blood must be somewhat increased through these trunks. But the stream might be *so* augmented that the amount of attainable velocity would not suffice for the discharge of the blood at a rate co-equal with that of its influx. The blood would consequently *overflow* the channels leading to them. It is under these conditions that an unusually large number of veins in the leg—*branches* of the saphenæ—become dilated, with danger of temporary embarrassment to the locomotor system, to disappear, however, as the cause subsides which rendered them prominent. Their repeated and long-continued dilatation makes this condition permanent, and leads to changes which ensue in confirmed varicosity.

In order, however, to connect these results with the muscular system and its functions, through the kindness of my friend Dr. Smiles, I had an opportunity of making a series of observations on the legs of a number of prisoners at the treadmill.

They are "worked" every alternate quarter of an hour from early morning; and I had the opportunity of making my inspection between one and two.

In fifty-two limbs I found the following results :—

In all, the saphenous trunk veins were, by the knee joint, visibly full, but more so in proportion to the number of distended branches and the degree of their repletion. They were filled in some instances without any concurrent repletion of the branches; but the branches were in no case filled without repletion of the trunks.

The following table refers to the repletion of the saphenous system, as it appeared to be associated with muscularity and constitutional strength :—

The veins are divided, according to the degree of repletion, into four classes. There were,—

1st. Very full, in five instances;
2nd. Moderately full, in nineteen;
3rd. Slightly so, in eleven;
4th. Not filled, in fifteen.

Of the 1st class—

One was a very muscular man;

Three were muscular; and

One was a thin, feeble man, with small muscles. He had been a vagrant.

Of the nineteen of the 2nd class—

One was very muscular;

Fifteen were muscular;

Two were pale and only moderately muscular; and

One was a pale, weak man, a tailor.

Of the 3rd class—

One was strong and muscular; he had only been in two days, whereas all the others had been in above a week.

One was muscular;

Two were strong-looking men, with *small* muscles;

Five were feeble, with moderate muscles; and

Two were pale, strumous-looking men.

Of the 4th class, in whom the veins were *not* filled,—

Two were feeble with moderate muscles;

Two were feeble with small muscles; one of these had a cough;

Five were feeble, skin anæmic (one had lepra), and with small muscles;

Three were strumous, with pale skin and moderately developed muscles; and

One was apparently a healthy man, with large muscles.

In a sailor, a man of middle height, with bulky, muscular legs, and yet with a good quantity of fat, the subcutaneous veins were enlarged throughout the leg.

These facts show that congestion of the saphenous system is closely associated with muscular activity; and not only so, but that there is to a certain extent—an extent sufficiently wide to serve as a basis for the view contended for—a direct ratio between the development of its vessels, but especially of certain primary and secondary branches, and that of the muscular masses with which they are connected. In young subjects I have observed a comparatively higher development of the muscles of the calf, than in those of the thigh, in cases of crural varicosity.

We have now reached a point at which we catch a glimpse of the mode in which I conceive saphenous dilatation, the first step towards varicosity, is for the most part primarily and directly induced; for, gathering up our conclusions so far, it would appear,—

1st. That the saphenous, in part *nutritive*, is in a still greater and more important respect an *appendage*—as a kind of reservoir, with slower moving currents—to the deep venous system, which belongs almost wholly to that department of the vascular system which may be termed *nutritive;* and thus, as Mr. Nunn observes, acts as a safeguard against the contingency of muscular congestion. *For in the venous system there are, potentially, no retrograde compensating currents as there are in the arterial.* Hence, *in all probability* dilatation of the *tegumentary venous radicles* indicates embarrassment of the deep trunk veins; whilst that of the *saphenous branches* points in like manner to obstruction of both superficial and deep trunk veins—of the first *directly*, of the second *indirectly*.

2nd. That in reference more particularly to the saphenous system, the dilatation of its vessels cannot be effected actively and solely by any force evolved by the various agencies directly concerned in carrying on the circulation, nor to the alleged disadvantages of the columnar pressure of the blood; but,—

3rd. That the relation of the saphenous to the muscular system indicates the source of a force, as well as the means of intensifying it, so far in excess of that which the vein walls can oppose to it, as satisfactorily to account for its production.

4th. That this result is attained in part,—

(*a*) Through a contingent disadvantage arising from the fact that the capacity of the saphenous trunks is greatly less than that of the sum of their branches; and for the rest, as I shall hope to make more clear hereafter,—

(*b*) Through the agency of certain foramina, of which *some* —*e. g.*, the saphenous, the femoral ring, and the triceps opening—regulate the size of the stream at the points where they preside over it, and, in case of overflow, intercept the surplus quantity of blood; whilst others—as the fascial and intramuscular—act as valves or barriers, and oppose its reflux.

5th. Blood so intercepted finds its way to, and accumulates in, the saphenous system; where its tension expends itself with varying results. The valved and powerful trunks are able to resist an amount that would dilate their unvalved and feebler branches.

LECTURE II.

I MUST this evening beg your attention to varicose disease as it is disclosed by the aid of the scalpel.

There are but few elaborate dissections of the vessels and structures concerned in cases of varicose disease on record, with the exception of those of M. Briquet and M. Verneuil. To these and their inferences I have already cursorily alluded; whilst the illustrations in works on morbid anatomy, as well as in our museums, are confined to inconsiderable portions of a vein, set forth, albeit with very questionable accuracy, as specimens of varicose saphenæ.

Cruveilhier, in his splendid work, limits its pathology to an account of the morbid anatomy of the superficial veins, and illustrates it by drawings of a thrombose saphena, with a loop branch in a varicose condition attached to it, in one instance; and by slight varicosity of saphenous branches, limited to their terminal segments, in another.

Mr. Hunter preserved a specimen (1736, Mus. Cat.) of "veins of a leg, dilated, very tortuous and elongated." One is stated to have a "thin layer of lymph upon a small part of its inner surface;" and at its upper part it is obstructed by a clot.

The integuments to which the veins are closely united are *hardened*, and the valves said to be shrivelled. In the same museum, contributed by Sir A. Cooper, there is a series of specimens (1733, &c.) from the limbs of a nobleman (Lord Liverpool), who had a swelling of the left leg and thigh, with a varicose state of the veins from the ankle to the groin, for several years before his death.* He died ultimately of apoplexy from, in Sir Henry Halford's opinion, obstruction to the circulation of the blood. His pulse was only forty-four in the minute. The whole length of the external iliac vein on the left side is obliterated by clot and greatly contracted. The femoral vein to the profunda is much dilated, and "in a *slight degree* varicose." The femoral artery is wavy; whilst its inner coat is "wrinkled and deeply seamed, and spotted with yellowish deposit." The right iliac vein is also obliterated by clot.

In other museums, specimens of so-called saphenous varicosity are also shown. In St. Bartholomew's, Sp. 77 is said to be the "saphena and its branches varicose." The saphena is, however, not varicose—there is a slight bulging of the coats in one part,—but the branches are; and they are said to contain relics of valves.

Mr. Nunn has recorded an interesting dissection.† The saphenous vein was found dilated and thickened from the groin to the knee. Below the knee the tortuosities commenced, and were found abundantly as far as the upper margin of an ulcer situated on the inner side, and a little below the middle of the leg. At

* The case is recorded by Sir H. Halford in *London Medical Gazette*, vol. x., p. 171.

† Nunn, op. cit.

the *seat* of the ulcer the vein lay in a bony channel formed by a thickened and ossified periosteum. Below it, the vein was not tortuous, but thickened, and its lining membrane roughened; whilst above, it was embedded in indurated tissue, to which it was not so closely adherent that the connections between the vein and the tissue could not be ruptured by gentle manipulation. The branches *communicating* with the deep veins were very much dilated, being about twice or three times their usual diameter; but their walls were so thin that they would scarcely bear the touch of the forceps. Throughout the whole length of the vessel the valves were found atrophied. The cutis was infiltrated with lymph, and the divided venous radicles appeared to be obstructed by brownish clotted blood.

Mr. Gross[*] relates a rare form. During life there had been great dilatation and tortuosity of the superficial veins of the leg from the base of the toes to the knee, attended with pulsation and a *bruet de souffle*, with a deep foul ulcer on the dorsum of the foot. The limb was amputated, and the patient (æt. forty-three) died of pyæmia on the twenty-eighth day after the operation. This person was born with a nævus at the back of the foot; at the age of three the foot began to swell; the superficial vessels then became enlarged and tortuous, and the temperature of the foot elevated. Six years after, an ulcer formed near the web of the great toe, and subsequently others,—one, the largest, over the inner malleolus.

All the superficial veins were found, on post-mortem examination, varicose to a striking extent, as were also the deep veins, but to a lesser degree. The internal saphena was three-

[*] *American Journal of the Medical Sciences*, No. 84, New Series.

fourths of an inch in diameter; its course was tortuous, and its coats were greatly thickened. The veins emptying themselves into it were much enlarged and twisted, and formed an intricate plexus over the dorsum of the foot, and anterior and internal aspects of the leg. On the plantar surface of the foot the enlarged veins and arteries formed a most intricate network, or rather, a confused mass of injecting material. (The limb had been injected.) The external saphenous vein was slightly enlarged, and accompanied by several smaller varicose veins. The posterior tibial artery was nearly as large as the femoral, and its satellites were twisted and nodulated.

This remarkable case shows, first, the influence of the heart's action in producing dilatation of the venous system when its power is not controlled by the intervention of the capillary system; and second, the difference between the results, as these are met with in the saphenous *trunks* and their *branches* respectively.

Mr. Canton relates a case[*] of a woman, æt. forty-nine, who died suddenly in consequence of the rupture of a varicose vein. There was no ulcer, and the aperture from which the blood flowed was a mere pin-hole. The veins of both legs were varicose, but of these the smaller were more extensively implicated than the larger. The disease commenced at the age of ten.

On examining the veins Mr. Canton remarked " how slight the tortuosity and enlargement of the saphena major was in comparison with these conditions in the smaller neighbouring tributary veins." A drawing is given in which the saphena is seen to be almost straight,—at most but a little wavy.

These few dissections reveal some important points;—viz.,

[*] *Lancet*, 1859.

the special character of the veins that become varicose, as shown in Mr. Nunn's dissection ; the thickening of the saphenous trunks, and their obvious freedom from an extreme degree of tortuosity and convolution, without regular thickening, which is implied in Mr. Gross's, and especially noted in Mr. Canton's case.

I must now give a cursory notice of some dissections recorded by M. Verneuil.

First, a case in which M. Verneuil accidentally discovered some small subcutaneous vessels at the inner and middle part of the thigh, slightly dilated and serpentiform. On pursuing them he found that these vessels communicated by a perforating branch with a muscular branch of the vastus externus, which, on being traced, was found amongst the muscular fibres, greatly dilated and beaded.

In another the intra-muscular veins were very numerous, enormously dilated, anastomosed freely, and completely furrowed out the muscular tissue in which they were situated. The soleus enclosed eight or ten large vessels, taking a direction parallel to its axis, averaging each in size that of a large goose-quill. They formed altogether a fourth of the volume of the muscle;—the muscle, in short, had become a sort of vascular sponge.

In a still more remarkable case M. Verneuil found, on the left limb of a woman, æt. fifty, a large tortuous vein which, rising in the upper and outer part of the leg, took a somewhat tortuous course around the back of the knee, and ultimately joined the internal saphena about the middle of the thigh. This was the only evidence of varicosity in both limbs. They were both injected. The varicose vein was found to communicate with seven or eight perforating branches which traversed the tibialis anticus and extensor com. digitorum muscles in the direction towards their origin. The intra-muscular veins of the gastrocnemeius and soleus were enormously dilated. The latter muscle was almost entirely furrowed out by these vessels. In the right limb, without any external manifestations, the size and disposition of the vessels were nearly similar in all respects.

One more case I must allude to,—that of a woman who had slightly developed varices which covered the right limb before and behind, from the knee to the ankle. In the left only a few tortuous vessels could be distinguished, owing to a remarkable thickening of the integument. Although small, these tortuous vessels communicated by large perforating branches with the deep in*ter*- and in*tra*-muscular veins. In both the deep veins of the calf were greatly dilated, and apparently to about the same degree.

These dissections led M. Verneuil to contend for the intra-muscular origin of varicosity, to which I have already referred.

I come now to some dissections which I have had the opportunities of making.

Tegumentary varicosity.

Dissection 1. A woman æt. fifty-six, left leg.

Small arborescent and linear varicose ramuscles were observed over the inner edges of the foot and malleolus, on the inside and outside of the leg, behind the popliteal space, above the patella, and on the outer side of the thigh. In some parts a patch of these varicose vessels were observed to centre in a large purple vein immediately below the skin which suddenly disappeared. These little vessels were beaded but not discoloured; and were found, on careful examination, to be formed into groups, which gathered their radicles from the confines of the capillary system, and themselves converged into a branch just larger than themselves. This branch formed as usual one of a series, progressively enlarging until it reached the trunk. The branches which intervened between the varicose vessels and the trunk were severely congested and convolute (Pl. III., fig. 2), but increasingly so in the direction from the trunk to the radicles; but they were not in the slightest degree varicose, nor could I find a vessel in which there was any obvious tendency to morbid action in the coats beyond the changes which I have mentioned. As they increased in size, so they became gradually less convolute, and left the higher regions of the skin in order to reach the trunk.

The internal saphena was firm and healthy throughout. It communicated with the plantar vein by a small branch, received the long dorsal vein just above the ankle joint, and, *after* receiving some very tortuous and bloated branches, began halfway up the leg to be filled with blood. From that point, as it received more and more tributaries, so it became more and more gorged to its termination. The external was represented by a very small empty and inefficient vessel. One of the venæ comites of the posterior tibial was distended from above the ankle, and became more and more so to its termination. The popliteal and femoral veins were enormously filled to the femoral ring. The muscles were somewhat redder than usual. The principal arteries were thickened.

The blood could not be returned into the external iliac by raising the limb, nor by urging it forwards from the trunks below. I therefore divided the ring upon a director, and with very obvious but still only partial relief to the vessels below. I could not carry my examination further.

It seemed to realize the view in relation to *tegumentary* varicosity already advanced, that it is associated with systemic *venous* embarrassment.

Dissection 2.

A fat man æt. forty-six, a cab-driver, died suddenly, and was brought to the Great Northern Hospital.

He had tegumentary varicosity in both lower extremities, but rather more severely in the skin of the left leg. I examined both, for which the description of one suffices.

The varicose patches were distributed over the limb, but principally along a broad track of skin which overlay the saphenous trunks, and especially on the site of a dusky or bronzed and soft or doughy portion of skin which extended from the calf to the ankle, and thence backwards to the heel. They were also distributed along the edges of the foot, on the outer side of the leg and thigh, and in the popliteal space. They were, for the most part, of the arborescent form, and made up of the smallest venous ramifications, variously beaded and convolute, and coloured—some of a bright pink, others of a purplish hue. The blood could be pressed out of them readily, and with it the colour disappeared, but it returned on remitting the pressure. Here and there a bloated segment of a large vein was noticeable at the point of convergence of a patch of these vessels.

On examining the body the left ventricle was found to be excessively hypertrophied, the aortic valves imperfect, and the coats of the aorta atheromatous and, in parts, earthy. The right side of the heart and lungs, with the systemic veins, were distended with dark fluid blood. The fat layer of the lower limb was very thick. The saphenous veins, to their minutest tegumentary ramifications, were gorged. The subcutaneous veins next in size to their varicose tributaries were severely convolute or serpentiform, and so were the successive branches, but in gradually lessening degree, to the trunk veins. Here the convolution ceased, but the repletion still existed.

In the dusky portions of the skin the venous congestion was enormous, from the aggregate capacity of its convoluted smaller veins. The blood poured out in a full stream. But below the ankle the congestion was comparatively trifling in amount. All the anastomosing channels between the veins of the foot and the deep veins were intensely congested and somewhat dilated, as well as the lower perforating vein, which passed through the soleus. The posterior and anterior tibial veins were healthy, and contained but comparatively little blood; whilst the popliteal, femoral, iliac, and cava to the heart were severely gorged, but not above their usual size. The last portion of the saphena was dilated to the size of the femoral. The fibres of the cribriform fascia seem to have yielded somewhat to the outward pressure of the current; but they still indented the coats of the vein (*a*, Pl. III., fig. 3). The arteries of the limb were of their usual size, but hypertrophied. The dusky portion of skin seemed to consist of an increase of the venous at the expense of its dermoid elements, with a superabundance of pigment in what remained of the upper layers.

The varicosity in this case was clearly associated with saphenous embarrassment, together with a like condition of the systemic veins and thoracic organs; and seems to support the views in relation to the pathogenetical origin of this form of varicosity which have been already alluded to. The triceps muscle in this case was more congested than other muscular structures.

Dissection 3 (Pl. II., fig. 4). *Mixed varicosity, with dilatation and thrombus of the deep veins.*

The right leg of a woman, with tegumentary varicosity on both its outer and inner aspects. She had borne children. I could learn nothing about her beside.

The varicose patches emptied themselves into slightly tortuous and dilated veins, which lapsed into their normal condition before they reached the trunks. A varicose vein (a) ran up behind the inner malleolus from a perforating saphenous radicle on the heel to the upper perforating vein (b). Its varicosity ceased as it entered this vein close to its fascial orifice and below a valve. A vein (c) with tortuous tributaries connected the external saphena below its orifice with the internal saphena below the knee joint. A branch (d) entered the lower perforating vein.

The deep veins were excessively dilated and full of blood, especially the popliteal and femoral; but their intra-muscular branches were healthy.

One of the posterior tibial (c, fig. 5) as well as of the peroneal venæ comites had become contracted; their coats were thin; and, with their tributary veins, they were filled with clot; whilst some of the tributaries of the *open* veins were also filled with old clot, but only as far as the first branches from their orifices. These clots were for the most part rounded off as they protruded into the cavities of the trunk veins (*a*, fig. 5); whilst from one, some tough, wavy tendrils depended (*b*). A thrombous mass (fig. 6) was found at the spur formed by the junction of the peroneal with the posterior tibial vena comes with branches into both vessels. This mass was made up of small compact emboli, some of which were yet loose, as though they had recently been brought up by the current.

On examining the blood which poured from the vessels of the heel, minute embolical particles, evidently not of recent formation, were found floating in it, which so closely resembled that of the emboli found above, that their identity could not be doubted.

The arteries were large, and their coats hypertrophied.

In this case also the tegumentary varicosity appeared to be associated with remora of the deep trunk veins; whilst the varicose branches bridged over the thrombous obstruction in the deep vein trunks. I cannot venture any hypothesis as to the clots.

In all these cases with deep trunk repletion, the expenditure of any contingent blood tension begins, not in the veins which connect the muscular system with its saphenous receptacles, but in the venous confines of the tegumentary nutritive system, from whence it proceeds in the direction of the trunks, superficial as well as deep; influenced in this respect, in all probability, by the sources of such repletion; *i. e.*, as these may have existed either in advance of or behind its seat; in the deep systemic veins or in the muscular circulation ;—or partly in both.

Subtegumentary varicosity.

Dissection 4. *A varicose loop vein.*

The vein proved to be, as in Cruveilhier's case, a loop from the internal saphena formed of the last portion of the long dorsal (Pl. II., fig. 2) and a short cross branch. Its orifices were very narrow, and the vein itself contained a beaded thread of fibrin which was loose in its channel.

The saphena was thickened, and its limitary membrane plicated in the longitudinal direction, and of a pinkish hue. At its upper part a varix had formed through destruction of the inner and middle coats of the vein between the cusps of a valve. The sac was lined by a layer of yellowish, tough, fibrinous material, and filled by old clot. Across its mouth a thick cord passed (somewhat in the form of a handle to a basket), composed of fibrin far advanced in organization, which had become united with the edge of the sac (Pl. V., fig. 3). For this purpose the limitary membrane had been removed over spaces corresponding to the size of the somewhat flattened-out ends of the cord; thus forming depressions, the surfaces of which were highly vascular. Into these the ends of the cord dipped, and to the opposing tissues they were attached by delicate areolar tissue. There was no disease in other vessels.

Dissection 5. *Varicose loop of internal saphena.*

This case was nearly a repetition of the former, with the exception that there was neither varix, nor fibrinous deposit within its cavity. The varicose vein was extensively sacculated and nodulated, and its convolutions were spirally disposed as well as serpentiform. The saphena was thickened, and its inner membrane plicated.

Dissection 6. *Subtegumentary varicosity;—asthenic form* (Pl. IV., fig. 1).

A fat woman, æt. fifty-seven, with, during life, large and feeble heart, œdematous legs, and varicose veins in both. The left leg was the more severely affected. A varicose vein was visible, running from the inner side of the knee over the thigh above the patella, and upwards along the course of the internal saphena as far as its termination. Portions of smaller branches, also varicose, were seen in different parts of the limb, as well as patches of skin affected with tegumentary varicosity.

The two limbs, after death, so nearly corresponded, that the description of one—the left—suffices for both.

The saphenous veins were healthy throughout, with the exception of being pale and somewhat thickened. The varicose veins commenced from healthy veins around the outer ankle, some of which anastomosed with the deep plantar, and others with branches of the external saphena. A large branch (a) passed upwards, and after crossing the tibia in the middle of the leg, joined the internal saphena by a short but straight terminal segment (b) below the knee. It received the two perforating veins (c, d) and a branch (e) from the outer side of the limb.

There were two other considerable varicose veins: one of these (f) arose from the saphena behind the knee joint; the other (g) from the outer side and back part of the calf, continuous with a perforating branch (h). These united and rejoined the trunk by a common branch near its termination. Of their tributary branches, some were varicose to their minutest ramifications, whilst others were healthy.

A considerable tegumentary vein (g) entered the terminal

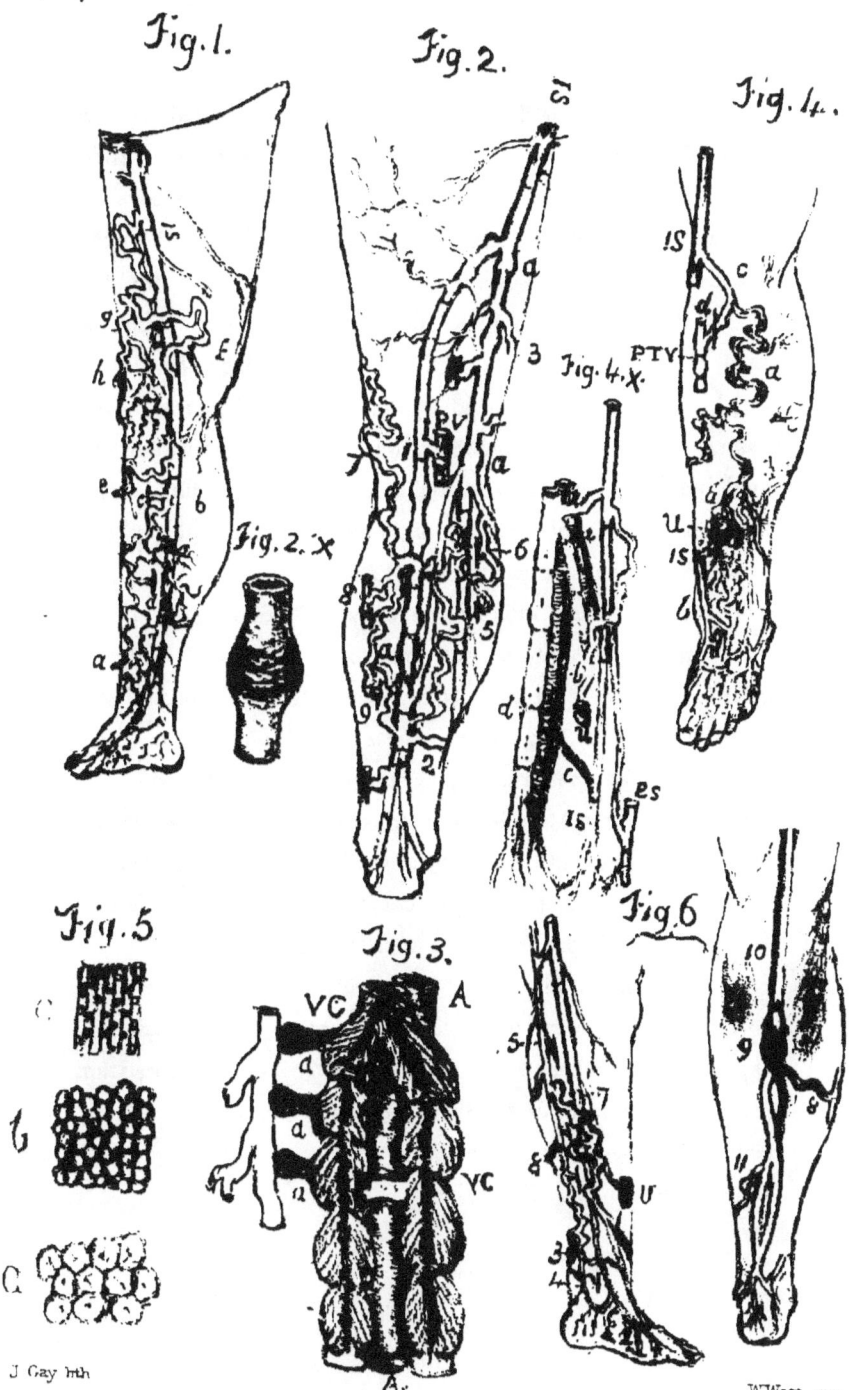

saphena, a short distance before the confluence of this with the femoral vein. Its orifice was above a valve and below a varix (*a*). A second dilated branch (*c*) communicated between the profunda and the saphena, but it was not varicose. Besides these there were other dilated and wavy veins which severally connected the saphena with the systemic trunks; viz., the pubic (*d*), the superficial epigastric (*e*), and external circumflex (*f*). The femoral (*g, g*) above the profunda to its termination was small and almost inefficient. In its course through the femoral canal it was closely environed by the cribriform fascia and other tissues. Below the profunda it was large and distended with fluid blood, as were also the popliteal and posterior tibial veins. A second varix (*i*) occupied the ultimate segment of the saphena.

The femoral artery was thickened and tortuous.

The blood in the saphena, on being pressed upwards in the direction of its orifice, was interrupted apparently by the cribriform fascia; for on relaxing the pressure, it regurgitated as far as varix *a*, but not into the other tributary veins. On cutting through the cribriform tissue where it surrounded this vein, the current became very considerably freed, but it was not *entirely* free until the tissues which enveloped the femoral vein at the termination of its course had been also divided, and the iliac released from its associate artery.

The varicose vein in this case bridged over the inefficient portion of the femoral; but the direct cause of the varicosity appeared to have been some hindrance to an overcharged stream just above the point of confluence between the saphena and the femoral; somewhat augmented, in all probability, by pressure on the femoral and iliac veins.

Dissections.

Dissection 8. *Varicosity with skin discoloration and intra-muscular dilatation* (Pl. II., fig. 4).

A woman, æt. forty-eight, had been a cook. On the right leg there was a broad patch of copper or bronze-coloured skin, extending from below the inner belly of the gastrocnemeius, downwards and forwards to the inner malleolus.

The internal saphena (1) was healthy; the external (2) thick and extremely bloated, from the outer edge of the foot to its foramen, and for some distance above. The deep aponeurosis was very dense, and presented a firm curved edge against the external saphena as it passed through it. Here (a) it suffered constriction; for on dividing the edge of the foramen, the blood, imprisoned below, shot up suddenly and rapidly into the popliteal, leaving the lower portion of vein empty. A small but old clot was firmly entangled in the valve at this point. The deep veins were healthy but congested to the triceps opening. The arteries were also healthy.

From the external saphena, immediately below its foramen, a large, bloated, and tortuous vein (3) ran across to the internal, which received, just before it entered that vein,—1st, a severely varicose branch (4), which took an erratic course upwards and inwards to the internal saphena behind the knee joint; 2nd, a large branch (5) which, after making a considerable curve below the point of its origin, traversed the soleus and terminated in the popliteal vein (7);—in its extra-muscular course it was exceedingly tortuous and dilated, but within the muscle (7) only slightly enlarged and wavy;—and, 3rd, a branch (6) from the saphenous loop below the inner malleolus.

These veins connected the external saphena on one side

with the posterior tibial and internal saphenous veins on the other, and thus bridged over a seat of obstruction which obviously accorded with the passage of the external saphena through its aponeurotic foramen. This vein appeared to have been overtaxed, and its place, in part, supplied by collateral branches which eventually became varicose.

Dissection 9. *Varicosity, with obstruction of the external iliac vein.*

In this case the external iliac vein had been almost entirely obliterated by the pressure of a cancerous tumour of the uterus.

The superficial veins were found to be largely developed, but there was no trace of varicosity below the upper part of the thigh. One small tributary of the internal saphena, which passed up behind the inner edge of the tibia to a perforating branch, was alone tortuous.

The femoral vein and its tributaries were small, with the exception of the profunda and its branches, which were large and slightly tortuous, in several instances, to their smallest visible twigs. The perforating branches belonging to this system were dilated in a part of their subaponeurotic course, but not beyond. Portions of the superficial pubic, external circumflex, and of some large veins belonging to the uterine plexus, were severely varicose, and appeared, together with the lumbar and superficial epigastric, to be the principal channels of communication between the common iliac and femoral veins.

The inner surface of the iliac vein along its compressed portion was irregularly raised, and figured with cords and bands, of which some were attached in their entire length, whilst others were unattached excepting at their extremities. It glistened like ordinary vein wall, and the raised tissues merged gradually into healthy vein below and above. On cutting it through vertically, the vein appeared as though it was much thickened; for the new could scarcely be distinguished from the normal structure. In the former, however, the microscope could discover nothing but wavy areolar with faint traces of yellow elastic tissue, and minute bodies which looked like *débris* of blood corpuscles. It was so closely adherent to the vein that their organic union could scarcely be doubted.

This dissection shows that it is not the tributaries of a vein below the seat of obstruction that become varicose, but branches which bridge over its site, and tie the first segments on either side.

Dissection 10. *Varicosity, with atrophied saphena and deep trunk dilatation* (Pl. II., fig. 5).

A woman, who had been a servant, and died at the age of forty-two, of phthisis, had varicose veins along the course of the saphena in the thigh, and others from the knee, extending behind to the external saphenous foramen, and below to the ankle. There was also considerable varicosity along the edges and on the dorsum of the foot.

In the leg the varicose veins brought the internal and external saphenæ into communication with each other, and with the subaponeurotic system, — 1st, by a straight but somewhat thickened vein (1) which sprang up behind the inner malleolus, where it communicated with the plantar veins; soon becoming varicose, it took a circuitous route across the front of the leg to the inner side, where it entered the internal saphena by different branches,—one about midway up, another below the knee joint; and 2nd, by a vein (2) which started from the external saphena below its foramen, and after passing directly upwards for a short distance, divided into two branches, one of which (3) rejoined its trunk in the popliteal space, whilst the other (4) helped to form a varicose tumour (5) on the inner side and just behind the knee, and also regained the saphena. Both straight and wavy branches (6) united the two varicose veins below the calf of the leg. There is no difficulty in recognising these veins in Plate I., fig. 1.

The veins in the thigh formed an extensive network along the course of the saphena. Healthy portions ran into others that were severely varicose, and diseased branches were connected by healthy cross-branches. These vessels together constituted a series of intricate loops, which, in the end, brought the middle and terminal portions of the saphena into communication with each other, but mainly through the medium of the pubic (9) and external circumflex (8) branches. One vein, for instance (7,7), took a long and erratic course, and after anastomosing with healthy as well as simply dilated segments, connected the upper part of the trunk with that which coasts the knee joint. This vein was severely varicose, and full almost to bursting up to the point at which it entered the saphena behind the knee. It communicated, by two healthy branches of considerable size, with the femoral vein.

The varicose tumour (3) consisted of a loop branch of the internal saphena, which had become much elongated, and, at the same time, convoluted and twisted in such a manner that the different convolutions formed a confused cluster, matted together by connective tissue. I succeeded, but not without difficulty, in disentangling the vessel.

The saphenæ were feeble and small, and contained but a very small quantity of blood, although the varicose veins were bloated throughout, as well as the deep trunk veins, as far as the triceps opening, and their branches to the nearest valve. The popliteal vein was dilated and somewhat thickened; whilst the femoral was almost empty.

There was no intra-muscular dilatation, with the exception of a considerable branch from one of the varicose veins which ran through the semi-membranosus to the popliteal vein. It is difficult in this case to allege any other cause for the varicosity than the ill-development or degeneracy of the saphenæ. The network of veins in the thigh appears to have been chiefly accessorial, and to have furnished important means of collateral circulation. The deep trunks to the triceps opening had been evidently overtasked; and from this cause the need of aid from those branches which became varicose seems to have arisen. I could gain no information of this person's history.

Dissection 11. *Varicosity, with skin discoloration and dilatation of deep trunk veins* (Pl. II., fig. 6).

The left leg of a man who died, at fifty-eight, of double pneumonia. Heart healthy; peritonitic effusion. There was some discoloration of the skin on the inner and lower part of the leg. In this case a portion of the external saphena (*a*) passed through the belly of the gastrocnemeius, accompanied by its nerve.

A varicose vein took its rise from the external saphena below its orifice, and after ascending a short distance divided into three branches. Of these, two (*c*, *d*) passed to the external belly of the gastrocnemeius, and thence to the inner posterior tibial venæ comites by the upper and middle perforating branches; whilst the other (*e*) diverged, and, crossing to the inner side of the leg, entered the long dorsal vein (L, D), just below its confluence with the internal saphena, and opposite its perforating branch.

The perforating veins passed through very narrow slits in the deep fascia, the edges of which slightly overlapped each other, and had to pursue somewhat zigzag courses between the aponeurotic fibres by which, in part, the soleus attaches itself to the bone. The movements of the muscle must have interfered with the current through these somewhat slender vessels. Their subaponeurotic portions were not dilated.

The deep trunk veins (fig. 7) were intensely distended to their intra-muscular tributaries; and these were much bloated for a short distance before emerging from, and slightly so within, their muscular connections.

The varicose branches were severely diseased. They had become narrow, thick, and twisted in some parts, and much attenuated through dilatation in others; and contained coagulum (fig. 8, *a*) unadherent to the walls, composed of threads of pale fibrine, with bulgings corresponding with the loculi. These bulgings, in contradistinction to the connecting threads, were covered with blood of a deep vermilion colour, which was readily washed away. I examined this coagulum very carefully with the able help of Dr. Reginald Southey. It was tough, and only tearable into fragments. These were made up of delicate fibres, separable again into minute bundles. On being teased out, these consisted of very delicate fibrillæ, laterally attached to each other, some of which, on still further separation, closely resembled finistrated tissue. Some few epithelia were found with it.

The tributaries (fig. 8, *b*) of the vein were also filled with coagulum, consisting of very fine fibrillar tissue. This was evidently advanced in point of age. In one specimen we found, on breaking it up, what appeared to be an agglomeration of casts of minute radicles, and a few fibres of what closely resembled unstriped muscular tissue.

It is not easy to explain the cause of the varicosity in this case. The varicose veins brought the lower portion of the external into communication with the internal saphena below the knee, and with the deep trunk veins, implying thereby some hindrance to the return of the blood through the upper portion of the external. This portion of the vein might have been inefficient, but collateral branches had sprung up which appeared to have been sufficiently compensatory. As there was no other obvious source of obstruction, the difficulties implied in the dilated state of the deep veins is the only clue to the varicosity that I can suggest. The varicose veins were, in all probability, undergoing a process of obliteration, there having been apparently no further need for them, at all events, arising out of any permanent or serious obstruction in the course of the venous system.

Dissection 12. *Varicosity, with dilatation of the deep trunk and intra-muscular veins.*

The right leg of the last body. The varicose veins and the superficial venous system generally, even to the unusual course of the external saphena, were nearly in correspondence with those on the left leg. The femoral, popliteal, and one of the posterior tibial veins halfway down the leg—that into which the intercommunicating veins passed (fig. 7)—were very greatly distended, together with their intra-muscular branches. The distention of the latter was scarcely perceptible in the smaller tributaries, but increased in direct ratio with their augmenting size. That of the posterior tibial continued to a valve below the orifice of a greatly distended branch. The segment to the next valve below was also dilated, but not nearly so much. Above each valve a muscular vein, also dilated to its first valve, emptied itself. Below these segments the vein and its branches were of their natural size.

This fact is interesting, as it indicates, in this case at least, the muscular origin of the varicosity; but, beyond this, two other points :—

1st. That the muscular tissue, as we saw in the experiments on dogs, contrives to empty itself of its venous blood, even with great stress on its larger branches; and 2nd, that the dilating force affects these veins in the direction *backwards* from the trunks.

Dissection 13. *Varicosity; thickened saphenæ, intra-muscular dilatation, thrombosis, and arterial disease* (Pl. III., fig. 5).

A woman, æt. fifty-six, died of erysipelas of the left leg. Both legs were varicose. I examined the right.

The coats of the saphenæ were slightly thickened, and the internal slightly wavy in its course along the lower half of the leg.

The varicose veins commenced in the leg by a vein (1) which took its rise from the external saphena, below its foramen. It divided into two branches (2, 3), which ascended a short distance at the back of the calf, and then diverged,—one taking a course through the aponeurosis and between the bellies of the gastrocnemeius, accompanied for a short distance by the external saphenous nerve, to the popliteal vein ; whilst the other, after taking a superficial course, joined the internal saphena below the knee. From these branches two others sprung (4, 5). The first passed along the inner side of the joint, and at the lower part of the thigh entered a kind of plexus of large veins, in part varicose (fig. 5, II). These veins took the course of the trunk with which they communicated, and after receiving some varicose tegumentary branches about the middle of the thigh, terminated by several branches, which entangled an inguinal gland in the last portion of the saphena. The second vein (5) passed upwards behind the joint to unite with the saphena in the middle of the thigh, sending a branch (6) to the popliteal on its way. Beside these, a considerable non-varicose branch (7) passed from the sural vein near the termination of its intra-muscular course to the saphena still higher up.

The femoral and popliteal veins were filled, but not greatly distended. The posterior and anterior tibial veins were large and bulbous between their valves, and filled, as well as their branches, with quite recently formed coagulum. The intra-muscular veins of the flexor com. digitorum were much dilated, and filled with old clot which adhered to their walls. They were also blood-stained through softening and partial desquamation of the limitary membrane. The trunk was bulbous, or beaded, in consequence of its constriction at one point by a band of muscular, and at another by tough aponeurotic, fibre adherent to the vein, which, with a nervous filament which also surrounded it, had to be divided in order to effect its release.

The posterior tibial artery was so encrusted, especially in the lower part of the leg, that its cavity was almost entirely obliterated. The disease in this vessel corresponded in situation with that of its comitial veins.

The varicose veins in this case bridged over the external saphenous and tricipital foramina, but at neither was there any very apparent source of hindrance to the current. Still the state of the trunk veins, deep as well as superficial, showed that the whole venous system had been overtaxed, and that these foramina were, as in the former case, *passively* answerable for the varicosity. Does the state of thé intra-muscular vein throw any light on the production of cramp?

Dissection 14. *Varicosity; thickened and dilated saphenæ; varices; deep trunk dilatation and general thrombosis* (Pl. IV., fig. 2).

This woman had been a domestic servant, and died of acute phlebitis, at the age of forty-five. She had renal disease. Both legs were varicose and œdematous. I will describe the state of the left, the worse of the two. Both saphenæ (I S, E S) were, with the exception of the lower third of the internal, greatly dilated, and invested with a thick sheath of dense and excessively vascular areolar tissue. The middle coats were also vascular and hypertrophied. Their inner surfaces were bloodstained, and from them thin laminæ could readily be peeled. The external saphenous, *below its fascial foramen*, showed these changes considerably in excess of the internal, as well as of its subaponeurotic portion. These vessels also presented, at somewhat irregular intervals throughout their respective courses, limited circumferential, spindle-shaped dilatations or *varices* (*a, a, a;* fig. 2), some below the mouths of large branches, others above valve cusps, and others still where neither branches nor valves were to be seen.

In the supra-aponeurotic portion of the external saphenous there were nine of these varices; they were comparatively less frequent in the internal.

The external saphena anastomosed, by numerous and large branches, with both deep and superficial veins in the neighbourhood of the ankle. Behind the malleolus, and above the first of its varices, it received a large and tortuous branch from a peroneal vena comes (1); and from this point it held frequent and free communications with the internal saphena. It gave a very large but short branch (1) to the popliteal, and then continued its course, as it not unfrequently does, to the back of the thigh, gradually releasing itself from the deep fascia, to terminate in the internal saphena, about halfway up. In the last part of its course it received several tortuous tegumentary branches of considerable size, and maintained free communication with the trunk vein.

The internal saphena took its usual course, but enlarged immediately after its junction with the cross-vein connecting it with the external below its foramen (2). In the thigh it sent a large branch (3) to the femoral, which was tortuous to its fascial orifice, but not beyond.

Two varicose veins arose by a common branch (4) from the external saphena below its foramen. These passed up, one on either side, forming a loop, the extremities of which joined a second loop, formed of branches which took their origin in a similar manner from the trunk at a point between its foramen and the popliteal space. From the first loop a varicose branch passed through the gastrocnemeius to the posterior tibial vein (5); from the second, two branches issued, one of which (6) terminated in the superior perforating branch of the internal saphena, whilst the other (7) passed obliquely upwards to the back of the inner condyle of the femur, and thence to the front of the thigh, where it entered a network of veins—varicose in parts only,—which terminated by several large branches in the saphena at the latter part of its course. This vein communicated with the posterior tibial veins through the upper and middle perforating branches. The deep trunk veins to the popliteal were throughout enormously dilated, and their walls thickened; and as, at the line of attachment of their valve cusps, they did not yield to such an extent as did the intermediate portions, so they had a beaded appearance (Pl. IV., fig. 3). Still the dilatation at these points was sufficient to render the valves to a great extent inefficient. The branches of these veins were also dilated, and somewhat tortuous to their nearest valves. The *intra*-muscular portions were not at all dilated, but the *inter*-muscular portions of some communicating branches were dilated beneath, not above the fascia. The inner surfaces of the comitial trunk veins were blood-stained, and the tissues softened. The saphenæ throughout were so dilated that their valves had become partially inoperative; and were, as well as their varicose branches and the deep comitial veins with their branches to their first tributaries, filled with coagulum.

The coagulum was not alike in all these vessels. In the saphenous and deep trunk veins it very nearly corresponded: it differed widely from the clot in the varicose veins.

In the former it filled the cavity of the vein, and consisted of a kind of central column of shreddy—in all probability post-mortem—clot, enclosed within a parietal column of conoidal lamellæ, which had their bases below and attached to the walls of the vein, and their upper borders incorporated with the blood of the central column (Pl. IV., fig. 3). This clot also contained all the elements of the blood; but the lamellated portion had separated into coarse fibrillæ, showing that it was recent, but not quite so recent as the other portion. It had become moulded around the cusps of the valves whilst in a semi-open state. The branches of these veins were thrombose to the nearest branch, but not beyond (*a, a, a*). In the varicose veins the clot seemed formed of concentric lamellæ, which enclosed small clot pellets; and corresponded with the shape of, but did not entirely fill the vessel, except in one instance, in which the varicose portion of vein terminated below in a large healthy branch. The end of this clot had been exposed to the force of the current, and showed cross wrinkles from being flexed or bent by its upward pressure. In the narrower sinuosities of the vein the lamellæ were attached to the walls. The nuclei or pellets consisted of bundles of distinct but somewhat coarse fibrillæ from which all trace of blood-globules had disappeared. The lamellæ, on the other hand, were shaded in colour. The lighter coloured portions consisted of blood *débris*, some white corpuscles, and separated residuum of serum. They were evidently old and of gradual formation, and perhaps originated in locular emboli.

In this interesting case the morbid phenomena which belonged to the fatal attack of phlebitis must be separated from the varicosity. The state of the coagulum in the varicose veins renders it very probable that these veins had been obliterated for some time, and that the circulation had been carried on by others, especially by the supplemental saphena in the thigh and the dilated internal saphena; whilst the coagulum in the saphenæ and deep veins was recent, and its deposit most probably associated with the fatal attack of phlebitis. The limitary membrane was softened and blood-stained; but the stain was easily taken out, as I found at first somewhat accidentally, by enclosing a portion in a tight noose.

The varicosity obviously arose from obstruction at the external saphenous foramen. The *whole* venous system, with the exception of the branches of the deep trunk veins and the femoral, had been exposed to embarrassment from great and persistent blood-tension. There was no *intra*-muscular dilatation.

Dissections 15 and 16. *Varicosity, bronzed skin, ulcer, and diseased arteries.*

The leg of a man æt. forty-nine. Both legs were similarly conditioned. They were contracted in size between the calf of the leg and the ankle; the skin was thickened, inelastic, and bronzed, the colour being gradually intensified towards the central part. The two legs so nearly corresponded, that the account of one adequately represents the condition of the other.

On cutting into the diseased skin, a large quantity of dark fluid blood poured out, as well from the smallest veins as from large tributaries of the external saphena. Two varicose veins, which communicated with the posterior tibial, passed up the leg to join the external saphenous below its foramen. These veins, very varicose below, became less so in their course through the denser portion of skin, apparently through incorporation of their outer coats with its diseased elements.

The external saphena was diseased from the point where it became sub-aponeurotic, to within a short distance of its termination; and, with the last portions of the two varicose veins, was, in some segments partly, in others entirely, obliterated by clot. Above each of its valves (it had three) the vein had become dilated into a circumferential varix, and the cusps in one had been entangled and rendered useless by the clot. Its walls were, in parts, blood-stained.

The clot showed evidences of having been altogether a long time in the course of formation, and of having been deposited first in the higher part of the vein, at the situation of a valve, and then gradually throughout its entire length. The upper (the earliest) portion was found to be attached firmly to the vein, so much so that they were inseparable. It had contracted, and formed a kind of *raised figured* surface to the vein, by which its previously obstructed channel had become in great part restored. Below this, it seemed to fill the vein; but on closer examination, it was found also to have contracted, but not so much as the portion above, and to be adherent to the wall, not only in mass, but by chords and flattened bands. Another portion of the clot, still further down, was moulded by the vessel, and attached to it by a very fine, delicate film of tissue only, whilst a yet more distal and very recent portion was entirely free. The surfaces of the clot and the bands were glistening, and seemed continuous with that of the vein itself.

It was clear that this coagulum had been formed by repeated additions to its lower extremity, and that it had undergone changes, in accordance with its age, from mere clot to organized tissue.

In the uppermost loculus the vein had undergone brownish discoloration over an oval space, and was thinner. The colour was mottled by some white, apparently atheromatous matter beneath the limitary membrane. The change seemed to be associated with an atrophic process, by which all the elements of the vein, especially the muscular, had severely suffered.

The posterior and anterior tibial venæ comites were healthy, but the corresponding arteries were much encrusted with calcareous deposit. Their branches, however, did not participate in this change; but were small, and their coats thin and feeble.

In this case the varicosity was obviously due to the obliterated saphenous trunk.

Dissection 17. *Varicosity, bronzed and indurated skin, diseased arteries, and ulcer* (Pl. IV., fig. 4).

A man æt. fifty-five, with ulcer on both legs, in the midst of deeply bronzed and indurated skin; one behind the inner, the other behind the outer malleolus, with marks of old cicatrices close by.

From the denser portion of the bronzed tissue the fat had disappeared, and become supplanted by coarse fibrous tissue, the fibres of which ran with the long axis of the limb. When cut through in the direction of its fibres, points were observed to be scattered with tolerable regularity between them;—the mouths of small veins from which blood could be squeezed, but did not flow spontaneously (c fig. 5). Towards the surface of the ulcer the tissue became gradually more coarse and white, whilst on the surface itself it was softened and completely disorganized. Proceeding from the sore, the fat element gradually increased, until at length, at some distance, it assumed its normal character and proportions.

The internal saphena (I S) was thickened, with the exception of that portion of the vein which traversed the dense skin. In their course through this tissue both saphenæ were compressed; their channels had become considerably narrowed; and their inner coats friable and easily separable from the outer. This, on the contrary, had become exceedingly thick, compact in its structure, and closely adherent to the adjoining tissue.

The veins of the foot were tortuous, and those along the inner edge, varicose. These communicated freely with the deep veins in the region of the ankle joint (*b*); whilst the varicose veins of the leg—the long dorsal and its branches (*a, a*)—united distant portions of the saphenæ on the one side, and these with the posterior tibial vein (P T V) on the other.

The peroneal and the outer posterior tibial venæ comites were narrowed throughout, and their channels obstructed by old blood clot (*a, b*, fig. 4) which pervaded their tributaries to the first valves, as well as a large branch (*c*) by which the latter of these veins was connected with the internal saphena. The internal posterior tibial vena comes (*d*) was dilated, the femoral healthy. The posterior tibial and peroneal arteries were atheromatous, and in parts encrusted and much narrowed. Their coats were softened; and the inner easily separable from the middle coat.

It would appear in this case that obstruction to the saphenous trunks by the diseased skin had greatly aggravated the varicosity, if it had not given rise to the varicose veins *below* the ulcer, since these communicated with the saphena *above* through the deep trunks. As I shall attempt to show hereafter, the disease of the skin was due to the condition of the deep vessels.

Dissection 18. *Varicosity, with indurated and bronzed skin, and ulcer* (Pl. III., fig. 4).

The right leg of a man æt. fifty-two, with an ulcer situated in a dense and discoloured portion of skin in front of the leg over the long dorsal vein, which was varicose from the ankle to the knee.

Stellar groups of radicles were scattered over the outer side of the leg and front of the thigh. In the leg, varicose convolutions were seen on the inner side of the tibia, about three inches below the knee. The radicles of the internal saphena below the inner malleolus and along the edge of the foot were also varicose, some of which contained arterial-looking blood.

The internal saphenous vein was healthy; the external was highly developed, and had a varix at its terminal segment, filled with old clot.

The long dorsal vein of the foot, after giving a large branch to the internal saphena above the ankle joint, became varicose; below this point its branches were healthy. It ascended the leg, crossed the tibia, and joined the inner saphena below the knee; but at a short distance before its termination it received the upper perforating branch, and here its varicosity terminated. This branch was large and tortuous; and had formed a very considerable channel by which the deep communicated with the superficial system of veins. It was in course of being obliterated by coagulum. The ulcer had encroached on and penetrated the vein, which had bled during life.

The femoral, popliteal, and tibial veins were distended with blood; whilst the latter were considerably dilated as well. The varicose veins of the foot communicated principally with the deep veins, and with branches of the internal saphena.

The external saphenous vein had in this case been obstructed, and the deep veins with the varicose long dorsal had supplied the necessary collateral channels for transmitting the blood from the extremity of the limb. Here, too, I cannot but think the varicosity in the immediate vicinity of the ulcer had been aggravated by it.

Dissection 19. *Varicosity, with bronzed and hypertrophied skin, diseased artery, and ulcer.*

A man æt. fifty-two. The ulcer was situated on the inner side of the leg behind the malleolus, in the midst of the diseased skin which occupied a large portion of the circumference of the leg between the calf and ankle. There was a congeries of varicose veins below the ulcer, extending along the heel to the foot, the edges of which were streaked by veins in a like condition.

The varicosity extended along the long dorsal vein from the upper margin of the ulcer to its confluence with the upper perforating branch, just below its joining the saphena, and along the perforating branch to its fascial aperture, but not beyond. It was severe. As it passed through the diseased skin, the vein had suffered compression; its calibre was diminished; its outer coat had become incorporated with the surrounding dense tissue; whilst its inner coats were softened and apparently breaking up into shreds which partially obstructed the channel. The main tributary had been destroyed by the ulcer, and its mouth plugged by clot.

The inner saphena, otherwise healthy, had suffered severe compression by the contracted skin. The external was dilated, and constricted at its orifice. Some recent clot was adherent to its walls above the foramen.

There were large communicating vessels between the varicose branches below the ulcer and the deep veins, especially with the posterior tibial venæ comites. Of these, one was very much thickened and dilated, whilst another was contracted, and its inner coats were undergoing a process of degeneration. Their accompanying artery was extensively calcareous.

The adipose layer had been converted into hard fibrous tissue; but a quantity of fat had been deposited upon the gastrocnemeius muscles. There was no *intra*-muscular dilatation, the muscles themselves being remarkably pale and soft.

In this case, again, there was obvious constriction of the external saphenous vein at its foramen, with evidences of embarrassment to the deep circulation, both from arteries and veins.

Dissection 20. (Pl. IV., fig. 6). *Varicosity, with bronzed skin and cicatrix of an old ulcer, and œdema.*

A man æt. fifty-six. Both legs nearly in the same condition. The discoloration and condensation of the skin occupied the space on the inner aspect of the leg from the calf to the heel. The ulcer was above the malleolus.

During life a knotted and varicose vein was seen crossing the tibia about the middle of the leg. There were some patches of tegumentary varicosity scattered over different parts of the limb.

The arteries were healthy. The long dorsal vein (1), which received a varicose vein from the plantar below the malleolus (2), took its usual course. A short distance above the joint it was joined by a varicose branch (3), which communicated with the posterior tibial by the lower perforating branch (4). The vein, before small and empty, enlarged almost suddenly at this point, and became varicose immediately above as it received the short dorsal branch, which was also enlarged and distended. After crossing the saphena, with which it had direct communication, it pursued its course behind the trunk, to unite with it in the middle of the thigh. Below the knee it received the upper perforating branch (5), but ceased to be varicose at the point where it was joined by the branch (6) which collects the blood from the inner side of the heel. Here the dilatation of the vein was such as to constitute a distinct varix (7). This vein was also varicose in some part of its course, and communicated with the posterior tibial by the middle and lower perforating branches, and with the external saphena by the cross-branch (8) below its foramen. This last branch was filled with old clot.

The external saphena below the outer malleolus received several large branches which collected the blood from a vast network of bloated veno-capillary vessels about the heel. It was bifid or double—the two vessels having lateral communication,—until it reached its foramen, and was surrounded by a dense sheath. Immediately below this point one of the veins swelled into a varix (9), into which the thrombose vein (8) passed, and which was also in part thrombose. The clot, which was old, was continued into and occupied about half of the cylinder of the vein above (10), to within a very short distance of its confluence with the popliteal. The other of the two veins (11) received a varicose branch from the peroneal, and continued its course into the open portion of the vessel beyond the varix. Above the foramen the calibre of the external saphena must have been very large before it was partially blocked up. It was, indeed, difficult at first to say that the vessel was not originally bifid, and that the coagulum occupied one of the veins, and not a portion of the channel of a larger vein. The popliteal vein had some clot, moderately recent, closely adherent to a portion of its walls. The internal posterior tibial vena comes was filled with recent clot; the external was empty.

In this case there was obstruction at the saphenous foramen, followed, in all probability, by thrombus, the result of phlebitis. The cross-veins between the two saphenous systems became obliterated, and a collateral circulation opened between the posterior tibial and peroneal venæ comites and the internal saphena, which ended in varicosity of portions of the intercommunicating vessels.

Dissection 21. *Varicosity, indurated and bronzed skin, diseased arteries, and ulcer cicatrix.*

A man, æt. sixty, died of cancerous deposit on the peritoneum and ascites.

Two varicose veins appeared to spring into existence above the ulcer and in the dense surrounding skin,—the long dorsal vein and its branch. They coalesced behind the knee joint; and the resultant vein, after receiving a branch from the posterior tibial, which had traversed the muscles of the calf, terminated in the inner saphena in the thigh. The saphenous tributaries of the foot as well as the smaller veins beneath and around the malleolus were varicose, and communicated freely with the deep veins.

The saphenous trunks were narrowed and thickened as they passed through the dense ulcer tissue; and in its passage across the tibia the internal was embedded in the periosteum, which was greatly thickened, and had become closely consolidated with the superjacent textures. In other respects the vein was healthy. The external communicated with one of the posterior tibial veins by a large branch which passed through the tibialis posticus. This vein was obliterated by a clot, whilst its fellow was free and much dilated. The arterial trunks were atheromatous.

The varicose veins were narrowed in their course through the dense skin tissue, to which their coats had become so closely attached and assimilated in texture that their channels had become almost obliterated, and what remained of them appeared rather like channels wormed out of the skin, than vessels independent of it.

The crural varicosity in this case was, in all probability, due to external saphenous embarrassment; whilst the varicosity in the foot followed the hindrance to the blood through the internal saphena and its branches, occasioned by the skin induration.

Dissection 22. *Varicosity, with indurated and bronzed skin, and ulcer on both legs.*

Man æt. fifty-four. Dissection of the right leg.

The long dorsal vein of the leg was varicose from the point where it gives off the upper communicating branch to the confines of the dense skin in which the ulcer was embedded. Beneath the ulcer the vein was friable and contracted; and did not appear recently to have taken part in the circulation. Below the ulcer the vein was healthy, but communicated, by means of large and somewhat tortuous vessels, with the posterior tibial venæ comites, which were very much dilated and, with the popliteal vein, were very full of blood. The accompanying arteries were thin. The internal saphena was healthy, but thickened in its course along the thigh; the external was also thickened, and obliterated by old coagulum, both below and for some distance above its foramen.

The varicosity in this case was probably due to external saphenous embarrassment. The vein appears to have become subsequently in part obliterated by the invasion of its precincts by skin disease, thus necessitating the opening of communicating branches between its lower extremity and the deep veins. The clot in the external saphena probably preceded both affections.

Dissection 23. *Ulcer and bronzed skin, with varicosity and diseased artery.*

The leg of a man æt. forty-eight.

An ulcer behind the inner malleolus, surrounded by a patch of bronzed and indurated skin, which extended from the middle of the leg downwards to behind and below the inner malleolus, and terminating, as is often the case, by an abrupt line, which ran from just above the ankle joint in front, obliquely, backwards and downwards to the heel.

This was the first case which I examined in which I was struck with the fact that the lower line of demarcation between the discoloured and healthy skin represents the anatomical boundary, below which the superficial have free communication with the deep veins; whilst above it such inter-communication exists comparatively but in very scanty measure.

The long and short dorsal veins were varicose to the entrance of the former into the internal saphena below the knee. These veins disappeared as they lay beneath the discoloured skin, and were found, on examination, taking a straight course beneath the ulcer, where their coats were so extensively dis-

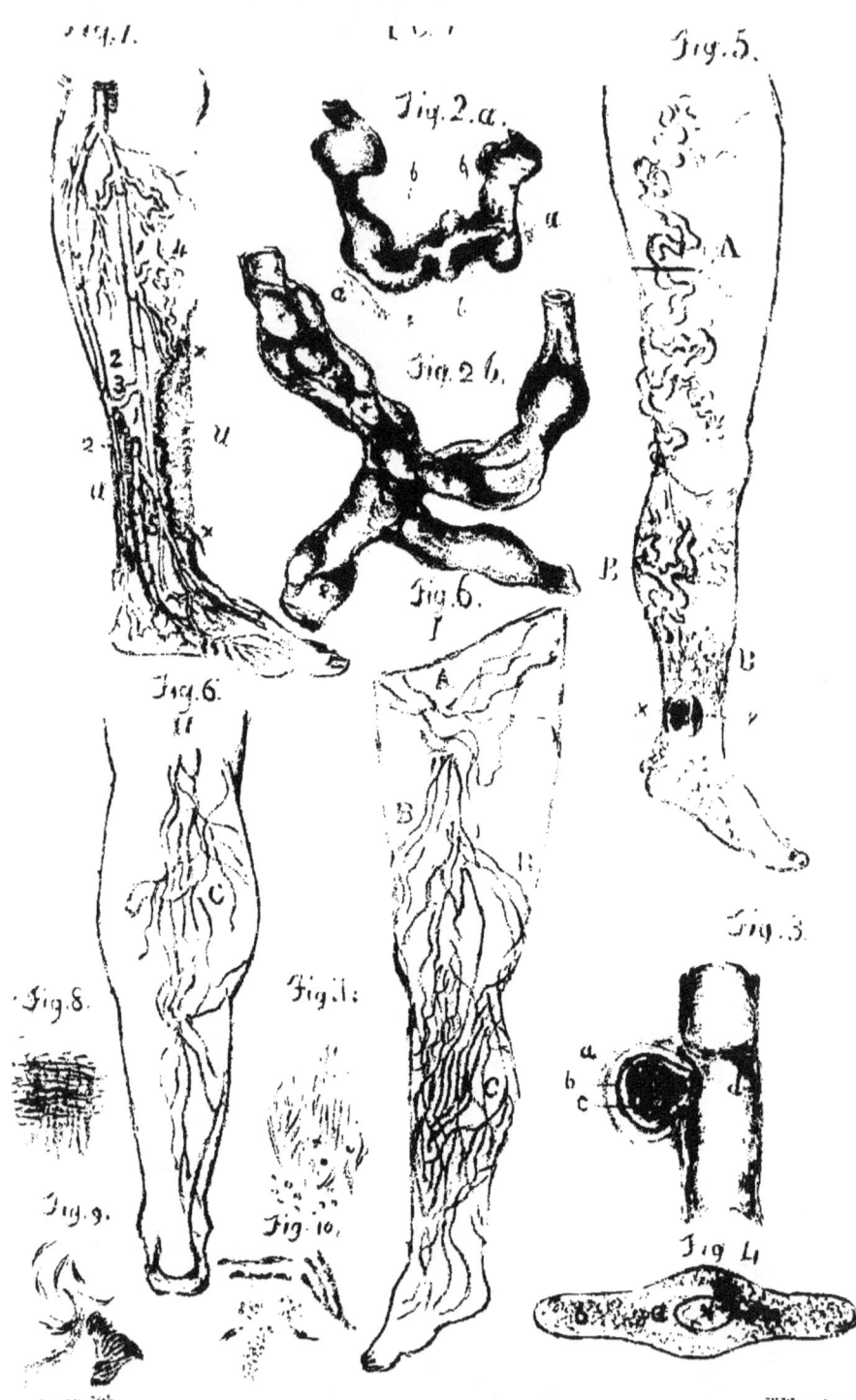

eased as to render them inefficient. At the lower border of the ulcer the long dorsal vein was dilated, and communicated by means of a large branch with an anterior tibial vena comes. The posterior tibial venæ comites were small; one was nearly obliterated by old clot; whilst the artery was extensively atheromatous. There was considerable venous congestion in the discoloured tissue.

In this case the crural varicosity, which was not extensive, appears to have been due to embarrassment of the external saphena. Its principal interest is its relation to the skin disease.

Dissection 24. *Varicosity, with skin induration, extensive arterial disease, and ulcer* (Pl. V., fig. 1).

A large ulcer occupied the lower two-thirds of the leg above the ankle joint, with the exception of a narrow portion of skin on the inner side occupying the space between the two saphenous trunks. Ulcers most usually occupy spaces *between* large veins.

It had existed many years in a fat woman of feeble habit. I removed the leg, and Mr. Hopgood injected it very successfully with wax, the patient having died of erysipelas.

The limb was exceedingly fat, but the fat for some distance around the ulcer was coarse, granular, and dense.

The ulcer had destroyed the integument to the deep fascia, and on the outer side had denuded and was eating away the tendon of the peroneus. The fascia itself was very much thickened, even to the portions that dipped down to form the sheaths of the tendons, and very vascular. The surface of the ulcer exhibited little plexuses of minutest arteries, from some of which the wax beneath the surface and near the edges there had escaped; whilst here and there were small collections of venous blood which had been recently effused.

The ulcer had destroyed everything in its way. The mouths of the long dorsal vein—which, with the tributaries of this vessel and the saphena in the foot, were the only vessels that were varicose—were open at each extremity (x, x). The saphenous trunks were unaffected by the ulcerative process. The internal (1) was healthy throughout, and lay along just outside the edge of the ulcer; whilst the external (2) lay just within the opposite border, and was compressed by the dense tissue through which it took its course. Its outer coat had become almost one with that tissue, and separated readily from the inner coats, which had undergone a process of softening. The vessel was, nevertheless, patent. A dilated and tortuous cross-branch (3) connected the two vessels below the external saphenous foramen. A varicose vein (4) connected the long dorsal near its termination with the saphena.

The popliteal artery was atheromatous. The anterior and posterior tibial and peroneal arteries were diseased throughout, with the exception of a small segment of the latter in the lower part of the leg. In their upper portions they were severely encrusted with calcareous matter, and their calibre thereby reduced to a third of its normal size; but their branches were of full size, and everywhere patent. Their ramifications were very numerous, so that the tissues were most minutely injected, and must have been during life exceedingly vascular. The portions of veins which accompanied the diseased arteries were so contracted as to be scarcely permeable; and their coats thickened, and of a dusky red hue. The nerves were very large and much injected; the popliteal and posterior tibial being translucent and soft,—apparently disposed to *ramolloisement*.

The arteries of the foot were healthy but small; its veins normal.

It is obvious that the ulcer in this case followed upon defect of nutrition consequent upon the disease of the main arteries. Still there had been no lack of blood; apparently it had been in excess, but had evidently not circulated through the destroyed skin in a manner consistent with the maintenance of its vitality. The varicosity was slight. Its direct cause must be doubtful.

These necropsies are not sufficient in number to form a basis for any general conclusions that are not fairly open to questioning and correction; but they establish facts by the aid of which we are enabled to examine and decide upon conclusions which have been arrived at by other observers; and they may at the same time be regarded as a kind of pioneering party, by which the way will have been so far cleared for further research into this interesting, though somewhat intricate, subject.

They bring before us, not only varicosity, but skin induration, and ulcer, with other concomitant morbid changes in sundry of the deeper parts of the limb, especially in its *arteries* and *veins*, as well as their contents. At first sight the varied morbid elements included in this category appear to defy an attempt at systematic arrangement, with the view of bringing together, in distinct groups, such as have a direct relation to each other as well as to some general result to which they mutually contribute: but on tabulating them, even somewhat roughly, some marked and interesting results are obtained, to which I shall now solicit your attention.

Of nineteen of these dissections there were nine in which there was simple external varicosity, ten in which varicosity was associated with discoloration of the skin and ulcer, and one in which there was varicosity with ulcer and skin induration, but no discoloration.

Now, although there was a variable amount of varicosity in these cases respectively, there was an absence of ulceration in a moiety of them; and in those in which ulceration co-existed, the severity of the ulcerative process did not correspond *pari passu* either with the extent of the varicosity, or the degree to which the veins were affected by it. On the other hand, the least amount of vein disease—that of a single vein—was met with in one limb with one of the worst forms of ulcer; whilst in another the most severe and wide-spread varicosity existed without any concurrent outward signs of disease.

Again, superficial varicosity, or varicosity of the saphenous branches, was found to be, in most of the cases, a complex disorder; and with its complications—those other changes of the skin, bronzing, induration, and ulcer—were in much more close correspondence than with the varicosity.

Now in two cases external varicosity existed alone; none of the *sub*-aponeurotic structures were in any wise diseased or disturbed. In other seven, with external varicosity, there were other complications, but no associated disease of the skin. These complications were as follows:—In two there were beaded strings of coagulum in the varicose veins; in two there was dilatation of the intra-muscular veins; in three the deep trunk veins were dilated; in five there were coagula in the deep trunk veins,—in two of these *obstructive;*—whilst in two others the arteries were *slightly* diseased.

This statement shows that, *with superficial varicosity, there are other serious lesions affecting both arteries and veins, deep and superficial*, which cannot be ignored in discussing its relation to associated diseases.

Now the several conditions of the various vessels in these cases were differently combined; and it is to these especial and individual groupings that the greatest amount of importance must be attached. Although coagula were found in the trunk veins in five out of the nine cases, these were, as I have said, *obstructive* in two only; and of these cases, in one only did the obstruction affect the *deep trunk* veins, the obstructed vein being in the other case but an *intra-muscular branch*. In one only *was there thrombose obstruction of the deep venous trunks with superficial varicosity;* and, be it further noticed, that in *this*, as in the other thrombose case, *as well as in one beside,—in which an old clot in the deep veins had ceased to be obstructive,—there was co-ordinate dilatation of other of the trunk comitial veins*.

It appears, then, that *in no one of these cases* in which varicosity existed alone was there coincident obstruction of deep trunk veins; although there were those especial signs of derangement of these vessels and their contents in most instances, and of the arteries as well in two, as would lead to the conclusion that the general circulation had been subject to very considerable and long-standing embarrassment.

In relation to *ulcer* the bearing of these remarks is this,—that varicosity may exist with mere *embarrassment of the whole vascular system of the limb, and yet without ulceration*.

I shall now go a step further, and give a summary of the condition of the vessels of the leg in ten cases, of which varicosity was

H

associated with bronzing and induration of the skin in one case, and with ulcer and discoloration in the other nine. Of the former I shall only remark that it arose from capillary congestion, and was associated with obstructive cardiac disease and great dilatation of the deep venous trunks.

In three of these, coagula were found in the trunk of the external saphenous veins. In six there was obstruction of the deep trunk veins by coagula; in three there was dilatation of these veins; and in six there was arterial disease, more or less severe.

I must again collate these cases, so as to present the distinct groupings of the several morbid phenomena.

In one case the external saphenous vein was obstructed with disease of the anterior and posterior tibial arteries. In two other instances in which the external saphenous veins were obstructed there was also obstruction of the deep trunk veins;—with, in one of these, dilatation of other deep trunks and arterial incompetency. In three others there was deep trunk obstruction, with like arterial incompetency; and, in one of these also, dilatation of other main comitial veins.

In one case, already referred to, in which a very large proportion of the integument of the leg had been eaten away by deep ulceration, all the main arteries were found coated with earthy matter, and had become thereby functionally obsolete. In this case, which was injected (Dissection 23), I was interested to observe the *unusual* vascularity of the remaining tissues as far as the arterial system was concerned. The diseased vessels sent off a large number of healthy branches, which supplied the tissues which the ulcer had spared, anastomosing freely with each other in the course of their ramifications.

Now, from the *second* series of dissections, it appears that, unlike the former, with venous obstruction, especially of trunk veins, superficial and deep, arterial incompetency was, in severe cases, found to be also associated. Contrasting this with the conclusion arrived at from the details of the first series, and remembering that there was no ulcer in the former, but ulceration with bronzing and induration in all but one of the latter cases, and that an unimportant one, it is impossible, I think, to avoid the further inference that *ulceration is not a direct consequence of varicosity, but of other conditions of the venous system with which varicosity is not unfrequently a complication, but without which neither one of the allied skin affections is met with,*—conditions which involve obstruction of the trunk veins, deep and superficial, either from impediments on the venous side, or incompetency on the arterial, or from both causes combined.

I infer, therefore, that pathologically, the doctrine of the "varicose ulcer" does not appear to "hold water;"—that, to reiterate my conclusions, ulceration, when it exists with varicosity but without other complication, is a coincidence and not a consequence of the vein disease; that, when associated with induration and bronzing of the skin, it is the direct result of serious obstruction of the venous trunks, and of this alone, whether associated with varicosity or not. The converse cannot, however, be alleged; viz., that serious embarrassment to the circulation through the trunk veins is *invariably* followed by these affections of the skin. The anomaly, which does not admit of explanation here, does not, however, invalidate the foregoing conclusions.

Having thus separated diseases which have been usually understood to have a constant and mutual relationship; viz., varicosity, skin induration, and ulcer; and endeavoured to show that they are, to a certain extent, independent of each other, and severally related to altogether different groups of morbid phenomena, I shall proceed to consider each separately, and somewhat in detail :—pathologically, etiologically, and therapeutically.

Of *varicosity*.—First, in regard to its general pathology.

It is somewhat difficult to detect, in the dead subject, that form of varicosity which I have termed "tegumentary," so that I cannot affirm that it did not co-exist with the ordinary varicose vein in a larger number of cases than I was led to suspect. In three limbs, however, its existence was very marked, and, as far as I could discover, there was no *varicosity* of the larger branches. The smallest venous radicles were the exclusive seat of the disease; the next and successive branches to the trunk were dilated and convolute; while in one, the trunk veins, in a second, the systemic veins, and in the third, the heart and thoracic organs generally, showed indisputable evidence of over-repletion. The *weight* of the embarrassment from such sources falls on the confines of the venous system; and its direct cause, in contradistinction to that of the ordinary varicose vein, lies in advance of, and not behind, the seat of the disease,—on the *venous*, and not on the *arterial* side. As the disease advances, so the little vessels worm out the dermis, and form a plexus, to the exclusion of almost every other element. Time forbids my making further reference to this part of my subject.

I must now refer to a few points in connection with the pathology of subtegumentary varicose disease; and shall begin with that of the varicose vein. Varicosity has been variously described by pathologists;—for instance, by Cruveilhier, as "lateral," "ampullary," or "multilocular" dilatation of a vein; as "dilatation with the formation of divisional septa," by Andral; as "irregular dilatation with hypertrophy and attenuation," by Briquet; and as "irregular dilatation attacking the veins at different parts," by Rokitanski. Either of these descriptions gives a fair account of such a vein so far as its general configuration and varying thickness go, but neither fully describes the morbid changes which especially characterize it.

A vein, in order to become varicose, elongates and slowly travels out of its ordinary track, making a gentle curve. Small dilatations or varices appear, like buds, corresponding, as I have already observed, for the most part with the points at which the vessel communicates through the fascia with the deep veins. It next breaks up into smaller curves or convolutions, and in parts twists upon itself. With these changes of general outline, others are taking place by which the vein loses its cylindrical shape. Fresh bulgings burst out along its course on the convexity of the convolutions, interrupted by segments which, on the other hand, contract and thicken; and thus, by the gradual exaggeration of these diverse conditions, the vessel ultimately loses all resemblance to a vein, excepting its being still tubular.

The migratory capacity of the vein is not so great in the immediate neighbourhood of the knee and ankle joints as elsewhere. In the former position, as the vein elongates, so it occasionally, as it were, doubles on itself, and forms that congeries of convolutions and tortuosities which makes up the already described "varicose tumour."

In its earlier stages, the outer surface of the vein, when removed from its bed, is smooth and irregularly covered with dense white areolar tissue, which, as the vessel becomes more and more convolute, crosses the segments diagonally, like bands, and accumulates in the interspaces between them. Ultimately it knits them firmly together; and, acquiring at the summit of each arch great density, forms with the contiguous portion of vein coat, with which it becomes here inseparably blended, the well-known *nodosity* (Pl. V., fig 2, *a*) often misrepresented as the remains of a valve, or as a phlebolithe. Should the convolution or tortuosity become unusually sharp, the nodosity may include an infolded portion of the vein coat, and be thereby made more firm and pronounced. There are, however, unevennesses which are met with in the course of a varicose vein which arise from other causes, as, unusual thickening of isolated portions of the vein coat and contained clot; but diseased valves are never, and phlebolithes seldom, the causes of nodosity.

The *channel* of a vein more advanced is followed with difficulty on account of its extremes of contraction and dilatation; and of the varied and sometimes acute angles and twists at which these contrasted portions unite with each other. In some instances a large pouch will lead abruptly into a narrow sinuosity by an orifice that will scarcely admit the smallest probe; and this as abruptly into a second pouch.

The *pouches* or *ampullæ* (Pl. V., fig 2, *b*), instead of being moulded within in conformity with their external configuration, are found, on opening the vessel, to be multilocular, from disarrangement of its several coats. The mouths of these secondary pouches or loculi are usually circular or oval; and are formed by the separation of the coarser fibres of the middle coat into a kind of web or network, through the meshes of which the inner coats are protruded.

The fibres of this network terminate in a thick band, constituted of the same texture which runs along the concave sides of the convolution; and with it the tissue forming the nodosity usually blends.

As the disease advances, so the vein wall becomes subject to degeneration of its histological elements; but I have examined no portion of a vein, however degenerated, in which traces of these elements have not been distinctly met with, with the exception of its epithelia.

The *limitary* membrane gradually loses its characteristic glistening surface, and becomes velvety and blood-stained. Moreover, it loses its elasticity, so that alterations of form become permanently retained. After a time the most delicate scrapings, submitted to the microscope cease to show epithelia, and but faint traces of finistrated membrane; but in their place blood débris and minute shreds of unstriped muscular and areolar tissue. These are sometimes found also with a clot that has been for some time adherent to the diseased wall. Thus the inner membrane softens, and in part and perhaps in some instances wholly desquamates.

The *contracted and thickened* parts of these veins result from the agglomeration of their coarser textures into a compact structure, in which the wavy, elastic, and muscular elements can be found, but not in their normal proportions.

In this extreme condition of the vessel there is still a channel, although, in some instances, barely permeable; so that it is not altogether one of complete uselessness.

The degrees of *attenuation* and *thickening* respectively, differ in these veins. In some, and especially in varicose veins in the leg, the attenuation is, of these changes, often the most marked. The vein, with nodosities and some slight corresponding constrictions, is almost wholly dilated; and forms, by the absorption of integument and the incorporation of its walls with the contiguous textures, a sulcus or a sinus through the skin rather than an independent vessel.

I now come to speak of the *valves* in varicose veins. There have been few writers on varicosity who have not thrown the onus of its production upon injured, diseased, or altogether destroyed valve cusps.

One of the latest and best authorities, Mr. Callender, says, " As the vein canal dilates, the valves, unless previously ruptured by violent and sudden muscular action, are unequal to close the passage; and being useless, they presently waste and are reduced to mere fibrous bands, or disappear altogether." *

Cruveilhier, however, in his commentary on Plate 74, asks, " What becomes of the valves"? " It was natural," he says in reply, " to suppose that the constrictions of the vein which separate the cellules answered to the basis of adhesion of the valve cusps, but the theory is not confirmed by the results of observation. The valves do not appear to enjoy 'un rôle bien prononcé' in the formation of varices, and they finish by *disappearing* in various veins." Pigeaux also stated that he had looked for valves in varicose veins, but had failed to find them.

* Op. cit., p. 313.

The reason I have already given. As a rule, veins that become varicose are destitute of valves; and the cords which have been supposed to testify to their existence in these veins, as I shall hereafter attempt to show, are referable to another source, viz., blood clot.

The *saphenous trunks* are prone to several morbid changes in connection with varicosity. I shall refer only to the most important.

These vessels are liable to *hypertrophy*, with or without *dilatation*, and a deviation from their respective courses into a series of *waves* or *gentle curves*. The muscular coat is generally the seat of this change, but it takes place without morbid deposit. This change, however, is very commonly associated with hypertrophy and opacity of the external or areolar envelope, and other abundant evidence of its having passed through severe attacks of acute inflammation. In two cases recorded, the evidences of such attacks—in one instance (Dissection 13) at the time of death, in the other (Dissection 19) at a period antecedent to it—were very clear.

These vessels are also prone to *atrophy*. In one instance (Dissection 10) it occurred with extensive varicosity and dilatation of the principal branches. The coats of the vein showed no evidence of organic change.

Varix is frequently met with in these trunks in two forms, viz., the *cylindrical* or *fusiform*, and the *lateral* or *circumscribed*.

In the *first* (Pl. IV., fig 2, *x*), the changes consist in general dilatation of the coats of the vein, with partial cleavage of the muscular. It is an interesting lesion, inasmuch as it can, as I have already remarked, be successfully imitated in the dead vessel by submitting it to the pressure of a column of mercury, varying from twelve to eighteen inches in height.* The walls at first dilate generally, then segmentally. Its permanency appears to be due to concurrent hypertrophy of their areolar elements.

Of the *second* form there are two varieties—the *true* and the *false*. The *former* is illustrated in the well-known saphenous varix, which consists in dilatation of the vein above its last valve. With attenuation of the several coats there is cleavage or separation of its muscular fibres. In the *latter* (Pl. V., fig 3), there is destruction of the inner and middle coats, the sac being formed of the outer coat. It is usually filled with clot, and lined by a layer of tough whitish fibrine, very similar to that which is met with in morbid specimens in the various museums, and is termed "coagulable lymph." Such a varix has been described in Dissection 5 (p. 65). It appears to have been occasioned by disease originally atheromatous, followed by softening and ulceration of the inner and middle tunics of the vessel. In Dissection 15 (p. 83), such changes appear to have been in progress in a portion of the popliteal vein of the size of a fourpenny piece.†

Varices of this class form in different parts of the vein. In one instance, its orifice was situated between the cusps of a valve; in a second, above; and in a third, in the museum of St. Bartholomew's, just below a valve.

* P. 36.
† Since the delivery of these lectures, Mr. Pick exhibited a like specimen at a meeting of the Pathological Society (see *Trans.*, vol. xviii., p. 56), and concurred in the opinion just given as to its pathology.

Although the precincts of these vessels are often invaded by ulcers, even to the tissues by which they are immediately bounded, the vessels themselves almost uniformly escape destruction. They do not, however, escape contamination; for the morbid processes by which their protection is effected—viz., thickening and degeneration of the periosteum fascia, or even bone, and the adjacent textures usually communicate themselves to the veins, and their coats undergo a series of concurrent changes, by which in the end their functions are gravely interfered with.

Their *valves* may become incompetent from dilatation of the vessel, or from the entanglement of a clot in their cusps; but are not often found torn. In one instance, however, I met with a cusp that appeared to have been torn away from its attachment on one side; and in another, a cusp perforated. I found it, however, very difficult, with limited time, to examine these appendages extensively.

Coagula are often found in these veins, but most frequently in the external saphena, in which I have met with them apparently in all stages, from that of very recent formation to that in which they had undergone a process of organization. Of these hereafter.

The *dilatation* of the *deep veins* may be either confined to one or more distinct intervalvular segments, or involve the whole of the vein, as well as the branches of the dilated segment to their nearest valves.

Their segments assume a bulbous shape, in consequence of the greater readiness of the vessel to dilate just above the valve. The *circumferential* is, indeed, the only form of *varix* that I have found in these veins, and represents their normal shape, exaggerated and rendered permanent by the loss of their elastic qualities. Like the saphenous varix, it can be imitated in a healthy vein by submitting its walls to the pressure of a column of mercury. With *general dilatation*, the edges of the valve cusps sometimes fail to meet. In one instance which I met with (Dissection 13), the valves were thus inefficient, and the vein was obliterated by old clot. The *limitary membrane*, too, is liable to undergo a process of softening and degeneration similar to that which, under like condition, takes place in the saphenous branches, and the vessel is thereby rendered prone to thrombosis. These veins are also liable to degenerate, through a process of inflammatory *thickening* with *softening* of the inner coats, and thus ultimately to obliteration, with or without clot deposit.

Intra-muscular dilatation—a condition of interest, on account of the importance which M. Verneuil, from his dissections, attaches to it—differs little, if at all, from the same lesion in other veins. It differs, however, as the vein is simply *intercommunicating* or *muscular*. Veins of the former class are often found *wavy* (Dissections 8, 10), with varicosity of the superficial veins; but I have not met with any morbid change in their coats indicating a tendency to disease analogous to that of their supra-aponeurotic correlatives. The *muscular* veins are subject to dilatation in direct ratio with their size and proximity to their trunks, as well as to degeneration of their inner tunics. The dilatation in the cases I have examined (Dissection 11) was irregular through the constriction of the vein by muscular and other fibres,—in one case, by a nervous filament. They are prone to varix, *circumscribed* and *diffused*. In one case, with slight external varicosity, I found a large diffused varix in the thigh of a sailor, æt. twenty-seven, who died worn out by stricture and peroneal fistula.

The sac was formed in part by the semi-membranosus muscle, which had been hollowed out through softening of its tissue, and for the remainder, by the femoral vein and artery, and intermuscular tissue. The latter was so thin that it gave way on the slightest handling. The sac contained two ounces of blood, which had been poured from a vein that had undergone destruction concurrently with the surrounding muscular tissue; its mouths were plugged. A very bloated vein, with a varix at its termination, entered the external saphena below its orifice, whilst another, also with a varix, entered it higher up. Both communicated at their opposite extremities with the internal saphena. The muscular veins of the gastrocnemeius were slightly, whilst the tributary veins of the external saphena were much, dilated.

The *valves* of the *deep* veins may also become partially or wholly incompetent, from perforation or laceration of their cusps—although I believe these accidents are rare—as well as from dilatation of the vein, or through fixation of the cusps by clot.

All the veins are prone to *thrombosis*, to which I shall now solicit your attention.

In all the examinations, save two, I found coagula in some parts of the venous system. They were not all thrombose; some were embolical. Nor were they all recent; many showed signs of having been deposited for shorter or longer periods. The coagula in Dissections 9, 11, 13, 15, 16, 18, 20, 21, 22 and 23 were formed in the situations in which they were found; whilst in Dissections 8 and 19 they had obviously been drifted to their present sites from remote parts.

Their *structure* and *composition* differed;—from the mere shreddy clot, with the elements of the blood in nearly their normal relation to each other, to the clot in which these elements had undergone changes which assimilated them to organized tissue.

They were found, too, in all parts of the venous system; in each in its turn.

In *form*, too, they differed. In Dissections 3 and 11 they were lying free in the veins—these were varicose,—and took the forms of the diseased vessels; in one instance, with bulbous enlargements opposite the orifices of their open branches. The examination of these particular clots made it appear as though they had been formed originally by concentric deposits around locular emboli, which had subsequently become united through the formation of intervening threads. The embolical clot in Dissection 19, as well as the clots which were found entangled in valves, consisted of rough fibrine with some blood corpuscles in a state of débris. That in Dissection 9 appeared to be made up of casts of veno-capillary vessels. I have met with such embolical particles in the blood of the smaller veins in portions of bronzed skin; it is, therefore, not unlikely that from this source contributions towards these embolical plugs are often made.

Again, coagula were found in branches, which presented clubbed extremities to the patent trunk vein, or fibrils, which dangled loosely within it. On the other hand, trunk veins were met with filled with clot; whilst their branches below their nearest valves were free. Shreddy, and apparently recent, clot was found adherent to the walls of the vein, from which the superficial film seemed to have been removed, leaving a red surface, with which the clot was in close adherence. Once more, coagulum was found in the same limb, Dissection 13, in different conditions, marking different periods of deposit.

In three instances clot was found to have become organized, and to have coalesced with the walls of the vein, giving to them a raised and figured surface. In one it occupied a considerable segment of the vein, and displayed from the lower to the higher parts of the vessel different conditions of advance from mere coagulum to organized tissue. The structure of these coagula has been described, p. 71.

I cannot omit allusion to a singular thrombus, of which I met with three instances in the course of my dissections. In each case it slipped out of the vein, and I was unable to track out its original seat. It was spindle shaped or fusiform (Pl. V., fig. 4), and consisted of two cysts—the first of the shape just described, the other (x) perfectly oval, of a smaller size, and situated in the centre of the former cyst. They consisted alike of delicate fibroid tissue, with smooth and glistening outer surfaces. The inner was the thickest and toughest of the two. The outer was filled by elements of the blood somewhat changed, and in a state of separation; the deeper layer (a) consisting of blood corpuscles and fibrinous material, whilst the outer (b) was manifestly the yellowish residuum of serum. The central cyst contained whitish amorphous tissue similar to that which, more compacted, constituted the cyst itself.

The coagula described appear to have originated either (1st) in *tardy circulation* through varicose channels, or through veins from which the stream had been diverted; (2nd) in *phlebitis;* (3rd) in *dilatation* of the vein walls with desquamative disease of its limitary membrane; or (4th) in *veno-capillary inflammation.*

Their relation to varicosity is a matter of interest as well as importance.

(1) As affecting the varicose veins: they would seem to be—either (*a*) remedial, as when they gradually fill the vein, and shut it off from the system at large without any abridgment of the means of perfect recurrent circulation; (*b*) idiopathic, and having no such desirable bearing upon the general system; or (*c*) of traumatic origin, and therefore also prejudicial.

(2) As affecting the other veins: they originate either in (*a*) the exclusion of a vein from the circle of the circulation through diversion of its current; or (*b*) in disease affecting the vein coats.

From a survey of the facts thus disclosed, it would appear that the formation of clot is generally *secondary* in relation to *uncomplicated* varicosity, but *often primary* to its *complications;* and that, consequently, varicose veins, when uncomplicated, must be regarded as accessories to a *new* adjustment of the venous system, often rendered more or less permanently necessary by subsequent clot deposits as well as by other sources of obstruction in the principal veins, superficial as well as deep. These considerations offer substantial reasons for regarding these veins as auxiliaries rather than as hostile to, or as "answering no good purpose" in, the circulation; and for the pertinacity with which they repeat themselves after attempts to compass their destruction. They explain, too, many other facts in the clinical history of varicosity, which would otherwise be meaningless.

But the formation of clot in large veins seriously compromises the venous circulation; and as, from what has been advanced, the system obviously admits of a very inadequate amount of relief from anastomosis, so its persistence would entail many difficulties. The clot does not generally, however, persist *obstructively*, as it began, except in those cases in which, as sometimes in varicose veins, it aids in restoring the venous system to functional order. The channels are often, to a great extent, restored, —not, however, as is alleged, by the breaking up of the clot or the solution of its elements,—but by its contraction, as these assume a higher condition, and eventually one of organic combination with the vein wall. The relics of these clots are, for the most part, cords and bands, with figured unevennesses of the inner surface of the vessel, to which I have already adverted.

I have referred to *barriers* which seem, from the condition of the blood and vessels, as exposed by dissection, to exert a considerable influence over the production of varicosity by regulating the stream in health, and impeding the superfluous quantity of blood when it is overcharged. The next diagram (Pl. V., fig. 6, I. II.) shows the respective tracks of varicose veins in twenty cases taken consecutively from my drawings and notes. It shows that varicosity is limited for the most part to a few, and these almost the same branches; and offers, at the same time, further evidence, to that already given, that the saphenous and tricipital foramina and femoral canal are chiefly concerned in those obstructive processes to which varicosity is due. From this diagram, the veins which are represented by the tracks are clearly divisible into three groups:—

The 1st (A) includes those which extend between the upper part of the saphena and the pelvic and lumbar veins, crossing the internal saphenous foramen and femoral canal.

The 2nd (B B), those which extend from the upper part of the leg to the upper part of the thigh, and bridge over the triceps opening; and

The 3rd (C), those which extend from the lower part of the leg to the saphena in its upper part and in the thigh, and to the popliteal vein, obviously bridging over the external saphenous foramen.

Of the three groups, the last comprises by far the greater number of lines; and in this respect favours the conclusion to which the facts and reasonings in the former part of these lectures point—viz., that the external saphenous foramen is by far the most common and efficient agent in the production of crural varicosity.

The next question relates to the *movement* of the blood in varicose channels.

Their varying morbid conditions would lead to the inference that, as far as their part in the circulation is concerned, they are from the outset at least functionally damaged; that their progress in this respect is from bad to worse; and that at length they become altogether inert. There are writers, and amongst them Pigeaux, who contend that such is not the case; and who have constructed a theory of varicosity on the basis of an alleged acceleration of the blood through these vessels, in consequence of an additional impulse given to it by capillary excitation; submitting, amongst other reasons for this view, the facts—1st, that hæmorrhage from varicose veins is always severe, and often quickly fatal; 2nd, that the blood has sometimes an arterial hue; and 3rd, that the veins have been known to pulsate. I can neither admit the theory, nor the explanation of the facts offered for its support. A varicose vein bleeds freely, simply because its blood has passed, in part, beyond the range of those forces which in healthy veins propel it onwards, and is therefore left to the consequences of its own gravitation. Invert the column, and the bleeding ceases. Whilst cases, in which these veins pulsate and contain arterial blood, are altogether distinct as regards both their causes and their pathological characteristics.

Varicose veins in pregnant women, after the third and up to the seventh month, if wounded, bleed furiously; but their condition is exceptional at this period, and by no means analogous to the state of such veins uninfluenced by this or like complications.

Let a varicose limb be raised from a state of recumbence, so that the foot shall attain a level above that of the pelvis, and the blood will in most instances immediately escape from all the superficial veins. It will as rapidly return on reversing the position of the limb in relation to the trunk, showing that the current in the veins, in which these simple phenomena show themselves, is to a degree subordinated to the ordinary laws of gravitation. But the blood does not in all cases escape from *all* the varicose veins when the limb is thus raised. Thus, in a number of limbs experimented upon, in which the varicosity was confined to the *leg*, the escape of the blood was complete in each; whilst in those cases in which the varicose veins extended to the saphena in the *thigh*, with the exception of one case—that of a strong lad of the age of nineteen—the blood did not escape from the *femoral* portions, although it still as rapidly disappeared from the *crural* veins, as well as from the portions of those varicose veins which entered the saphena near its termination. The blood did not only fail to escape, but it resisted in various degrees the influence of pressure so applied as to force the current onwards; and, on the remission of the pressure, it invariably returned with a rebound. In all cases the veins were refilled on the patient assuming the erect posture.

It is clear that in the upper part of the thigh there was in these cases a centripetal, though impeded current maintained by the ordinary resources of the circulation. In the *leg*, on the other hand, the blood could be poured backwards and forwards at will; but this did not arise entirely from the vitiated state of the vein coats: it arose in part, and in some limbs wholly, from the want of that assistance which in a normal condition these vessels derive from the *arterial* side of the system; for when emptied, they did not refill from that side at all in some instances, and in others but very tardily. In the cases in which the veins did not fill from the arterial side, the coats were extremely vitiated, and the varicosity was in most instances complicated with ulceration; whilst in those in which the stream returned in its normal course, the vitiation varied in degree much in correspondence with the rapidity with which the current was restored.

Pl. V., fig. 5, for instance, represents an aggravated case of varicosity, shown me at the Marylebone Infirmary. On elevating the leg above the level of the body in this case, the veins (B) became speedily emptied and remained flaccid. At A the blood was lessened in quantity, but it did not entirely disappear. On pressing it onwards by the finger towards the saphenous orifice, the blood could be almost entirely extruded from the vein; but on remitting the pressure, it returned in the direction of, and as far as, the knee. On pressing it from the vein (A) backwards towards the foot, the blood filled the *before emptied* veins of the leg; but on carrying the pressure slowly and methodically downwards along these veins as far as the ankle, they remained empty *behind the finger, and only refilled upon the blood being again forced into them from the large branches in the thigh, or being permitted to fall into them, as it readily did, by regurgitation.*

Such experiments—and I frequently made them with like results—show that, in cases of varicosity, there is no lack of channels by which the blood may escape from the limb, but a want of power on the part of the circulating system from various causes to utilize them. A normal arrangement of the blood-courses is supplanted by one that is abnormal and often persistent. But there is no actual stasis in varicose veins; for even when the vein is so vitiated that the blood takes a retrograde course, it still finds its way by devious courses to healthy veins, where it is brought again within the sphere of the normal circulation.

Hence it must be conceded that varicose veins, although depraved to such an extent as to have lost the qualities which in health distinguished them from simple inert tubes, are still useful so long as their continuity is not destroyed.

I must conclude this lecture with a summary of the conclusions at which we have arrived.

They are as follows:—

1st. That varicose disease of the lower limb includes a variety of morbid phenomena, from its simplest form, in which these are limited to varicosity of a saphenous branch with some change,—generally dilatation in the allied portion of the trunk vein,—to its more complex, in which they involve as well the veins of the deep or subaponeurotic region.

2nd. That these phenomena indicate as their direct *cause* the expenditure, upon the saphenous as well as the deep venous system, of a powerful force, which differently affects their several orders of vessels, and issues in a compulsory *re*arrangement of their blood-courses, on an abnormal type.

3rd. That the rearrangement or redistribution of these channels, at first temporary, may, and does often become permanent by corruption of their walls and appendages as well as by the formation of thrombi within them.

4th. That *sthenic* varicosity is a special disease of certain unvalved branches of the saphenæ which bring into direct communication with each other;—either (*a*) distant portions of the same trunk; (*b*) the saphenous trunks themselves; or (*c*) these and the deep trunk veins.

5th. That varicose disease originates in the branches of the saphenæ, from whence its course is to extend itself to the muscular system of veins.

6th. That the saphenous trunks are prone to phlebitis; the deep veins to dilatation, with organic changes affecting principally their inner and middle coats; whilst those other and more severe morbid changes which constitute varicosity are limited to those especial communicating branches of the saphenæ which have just been alluded to. As the latter form of disease is irretrievable, its natural "cure" is only effected by the diseased vein becoming plugged or otherwise obliterated, as the circulation becomes restored to its normal condition so as no longer to require its aid.

7th. That whilst the formation of clot within the varicose vein is often a salutary act, it is not so to be considered in the deep veins in cases of varicosity. The channels of veins, however, that become thus obstructed may, after a time, be efficiently restored by the organization of the clot.

8th. The *asthenic* variety differs from the *sthenic* inasmuch as, while in the latter the disease is usually confined to the main branches, in the former it invades the tributaries to the smallest ramifications. Fatty degeneration of the muscular coat forms, I have reason to believe, its distinctive pathological feature.

9th. Obstruction, direct or indirect, to the flow of blood through the trunk veins, is the immediate cause of varicose disease.

LECTURE III.

I HAVE to-night to conclude these lectures, but have much yet to say.

The question of *etiology* next presents itself. As I have already shown, every conceivable agency that can affect the venous circulation in the lower limb has been said to contribute more or less to the production of varicose disease. And, no doubt, with some good reason; but it remains to examine some of these agencies, and to assign to them the kind and amount of influence which they are severally capable of bringing to bear upon this result.

If our premises warrant the conclusion that *tegumentary* varicosity is due to embarrassment in the systemic venous circulation, whilst the *subtegumentary* form is due to venous embarrassment originating on the arterial side, it becomes comparatively easy to determine their respective causes, and to allot them severally to that form and condition of varicosity with which they are etiologically related.

They may be thus classified:—

1st. Those which are said to produce varicose discase by

acting on the veins in the centripetal direction, or in the direction of the current :—

(*a*) Muscular exercise, as influenced by occupation, habits, age, stature, and sex.

(*b*) Arterial and venous anastomosis.

These relate especially to subtegumentary varicosity.

2nd. Such as are supposed to have the same effect by acting centrifugally or in opposition to the stream; and are therefore related to the tegumentary form, and only subordinately to the ordinary varicose vein :—

(*a*) Gravitation of the blood in the veins;

(*b*) Valvular inefficiency;

(*c*) Abdominal and pelvic tumours;

(*d*) Utero-gestation;

(*e*) Pressure of left iliac artery;

(*f*) Impaction of colon;

(*g*) Hernia;

(*h*) Intra-thoracic disease.

3rd. Disorders of menstruation.

4th. Causes that may be said to be *predisposing*, which include amongst others all forms of exhausting disease; amongst which the following appear from clinical observation to be the most constant:—

(*a*) Heritage;

(*b*) Fatty degeneration of the coats of the vessel;

(*c*) General obesity;

(*a*) Fevers;

(*e*) Gout and rheumatism; and

(*f*) Chronic phlebitis, according to Rokitanski.

5th. Injuries.

Let us turn to the first series of causes, and see how far the *social* history of varicosity can assist us in determining them, and their amount of influence respectively.

It is a very old maxim that varicosity is most constantly met with in persons accustomed to occupations involving excessive exercise of the muscles of the lower limb, especially if it be prolonged and in the upright posture; more especially still, if the limb is simultaneously exposed to heat or moisture. Briquet made these observations; and they have been confirmed by P. Boyer, Sistach, Verneuil, and Longet. Thus Briquet found it most common amongst soldiers, printers, porters, hawkers, pedlars, waiters, woodcutters, and "nymphs of the pavé."

Of fifty-eight cases into which I made especial inquiry, the disease was met with most frequently in pressmen and in warehousemen; then, as in order given, in men given to athletic sports, in bookbinders, household servants, whitesmiths, ostlers, and only in one instance in a tailor. This man was, however, not a naturalized son of Crispin; for when off "the board," he was constantly at cricket and other field sports. In the middle classes, amongst men, I have met with many cases of varicosity from excessive devotion to athletic sports, and, for the most part, in tall men. For instance, in one it appeared severely after running and jumping in excess for a period of two years; following the hounds on foot induced it in another; whilst in a third, a man who indulged a like taste, but who, from injury to the great toe of one limb, threw the onus of his indulgence on the other, became soon debarred from it by the veins of the latter becoming severely affected. The act of standing *for a prolonged period* has been said to be more conducive to varicosity than that of walking or running, and I have certainly found the disease severe in men whose limbs have been exercised in this manner—viz., in printers' pressmen; whilst between riding and walking, the influence is greater on the side of the latter;—a fact authenticated by my friend Mr. Lawson, from his Crimean experience.

As, according to M. Briquet, varicosity was not met with in his time (1824) amongst the higher ranks of society, I was curious to ascertain, if possible, whether the remark is still of equal force. I therefore sought information on this point of professional friends, who are more cognizant of the disorders of high life than myself, and all concurred in the spirit of the following statement with which Sir William Fergusson favoured me. "Varicosity," says Sir William, "is as common, in proportion to numbers, in the upper classes as in the middle or lower, and is apparently more frequent in women than in men; but it occurs so often in men that the difference is of no practical importance."

This is somewhat remarkable testimony to the change which has taken place in society during the last few years. Young ladies may not be delighted at being overtaken with varicosity, and condemned by the matter-of-fact surgeon to rest and laced stockings. But the change is, nevertheless, a subject for modified gratulation. It betokens a transition from an age of indolence, ennui, and pallid cheeks to one of physical activity and cheeks of rosy hue; from the minuet to the polka and galop; from the carriage to the "Row" and the "Meet;" from Brighton Sands and Bath Waters to high latitudes and Alpine adventure. Surely this evidence of the spread of muscular Christianity is a happy omen for the future of old England, albeit at the cost of a little maidenly varicosity.

The left limb has been said by some authors—Richerand, Parent-Duchatelet, Blandin, P. Boyer, Dupuytren, and others—to be oftener and, for the most part, more severely affected by varicosity than the right. M. Briquet, to whose laborious paper I must again appeal, states, on the contrary, that the right limb is most frequently, if not most gravely, affected, and that in the proportion of 15 to 13. M. Verneuil collected 100 cases from various sources, and observed that of these, varicosity was bilateral in 30, on the right side in 36, and on the left in 34; in 12 living subjects bilateral varicosity existed in 8, unilateral, right, in 3; unilateral, left, in 1; whilst in 13 dissections the varicosity was double in 9, on the right side in 2, and on the left in 2. Taking 106 unilateral cases: in 56 it was on the right, whilst in the remaining 50 it was on the left side.

When the varicosity is *double*, says M. Verneuil, it is generally in excess on the right side; and injections of the vessels showed, not only that *traces* of varicosity were more frequent on that side, but that concurrent affections of the deep veins were also more advanced on that than on the opposite side.*

I took notes of 58 cases in succession. Of these 26 were in men, 32 in women. In 10 of the men the varicosity was in the right leg, in 4 it was in the left, whilst in 12 it was in both legs. In two cases the affection was most severe in the right, in the other two it was most severe in the left leg.

Of 32 women, in 5 the varicosity was in the right leg, 9 in the left, 18 in both legs.

Of 5 cases, in 3 the disease was most severe in the left limb.

* Gazette Hebdomadaire, Vol. II., 1855.

These numbers somewhat correspond with those both of M. Briquet and M. Verneuil. In the greater number of cases of *unilateral* varicosity, the disease affects the right limb; as well as in bilateral cases, when a difference exists.

There is, however, in this respect a difference as far as the sexes are concerned; for whilst in men the disease is most commonly *unilateral*, in women it is most frequently *bilateral*.

Of age.

Briquet says that varicosity rarely commences before puberty, the most common age being "between 30 and 40—that at which men accustom themselves to vigorous exercise, and women have usually borne children." Women often date varicosity to the period at which they begin to menstruate. It never begins in old persons. Out of 50 men below the age of 30, Briquet found 4 with varicose veins; of 30 women of the same age, only 1; of 60 men beyond the age of 30, 15 were varicose; of 93 women, 12; of 158 old men, there were 53 so afflicted; and of 392 old women, only 30.

M. Sistach (whose tables are exceedingly valuable) says varices are rare in infants, frequent in adolescence; they attain their maximum of frequency between the ages of 30 to 40, become progressively less frequent after that period, and tend to become stationary, and even to diminish in volume and severity, in old men.

I find that in France, from 1850 to 1859, "Les Conseils de Révision" exempted 41,325 *young* persons in consequence of their being affected either with varices or varicocele.

Amongst the French soldiery, 75 per 1,000 were affected between the ages of 20 and 30, and 194 per 1,000 from 30 to 40. Amongst the Zouaves, the proportion was 37 per 1,000 between 20 and 30, and 92 per 1,000 between 30 and 40.

In a Parliamentary return for 1861, out of 7,993 recruits for the English army, under the age of 25, 326 were rejected for varicosity; and of these numbers it is curious to observe that in England they were in the ratio of 27 per 1,000, in Scotland 54 per 1,000, and in Ireland 67 per 1,000.

In the militia, of 1,028 candidates who presented themselves for that service from 1860 to 1865 inclusively, my friend Mr. Childs informs me 35 were rejected for the same cause. Their ages varied from 21 to 33.

Of 74 consecutive cases which came under my own observation, including 28 males and 46 females, their ages were as follows, taking decennial periods:—58 were between 15 and 40 years of age; and the largest number, 25, between 20 and 30.

Males.	Females.	Ages.
6	8	10 to 20
10	15	20 ,, 30
4	15	30 ,, 40
6	6	40 ,, 50
	2	50 ,, 60

On the whole, it may be said that the first traces of this malady are more frequently seen in women between the ages of 30 and 40; whilst in men it is more constant in the preceding decenniad. I have seen lads of the age of 14 with varicose veins.

It follows that varicosity is, at all events to a great extent, a disease of the young, and that the ages at which it most constantly develops itself are those characterized by functional and physical activity.

According to Briquet, men are more liable to varicose disease than women. Of 258 men, 71 were so affected; of 493 women, only 42. In a number of consecutive cases, I found that the comparative frequency was reversed; but the question is of little importance, since it must be determined by districts and classes of society, and it would be difficult to obtain a return which would represent a general result.

I have observed it, on the whole, as frequent and as aggravated in the single as in married women.

Now the *tegumentary* form, which is most frequent by far in women, is developed at an earlier period; and, from my records, it has been as frequently observed in persons between the 13th and 20th year, as in persons of more advanced age.

Of arterial and venous anastomoses, as a cause of varicosity, I have already spoken,—see page 114.

The next series, the *second*, includes causes which are supposed by Begin, Briquet, Nellaton, and Bouillaud to produce varicosity by obstructing or retarding the flow of the blood through the systemic veins.

The *gravitation* of the *blood* can by no possibility have the effect of dilating the intercommunicating branches of the saphena, so long as the saphenous valves, as well as those of the deep veins, are so far efficient as to resist the regurgitation of the blood. And there have been no instances that I know of, in which varicosity was shown to have been *preceded* by valvular incapacity (see p. 38).

Neither is direct *obstruction*, either of the *vena cava* or of the *iliac veins*, a cause of varicosity in the lower limb. Varicosity *may* follow such an occurrence, but it is abdominal or pelvic, and not crural. Through Mr. Fuller's kindness, I dissected a case at the Marylebone Infirmary, in which there was perfect obstruction of the iliac vein and first portion of the cava: the superficial abdominal and thoracic veins, which formed collateral channels of communication between the last branches of the femoral and the internal mammary and axillary veins, were enormously dilated and varicose; but the veins of the lower extremities were without any participation in the disturbance.

Of 33 cases of *ovarian* and *pelvic* tumour with subtegumentary varicosity, I found it severe in 3, and in a milder form in 5 others, whilst in almost every case the tegumentary radicles were more or less severely affected.

On inquiry into each of these 33 cases, I found good reasons for believing that the varicosity had existed long prior to the growths, and in only one—a case, apparently, of malignant disease—did the state of the veins appear to be materially aggravated thereby.

In one case at present under observation—a Miss H., ætat. 44—varicosity exists in both legs; there is an ovarian tumour on the left side, which has acquired a considerable size. As this has advanced, however, the condition of the varicose veins has improved; and, as far as these are concerned, her capability of exercise is greater than it has been for some years.

Mr. Spencer Wells informs me, as the result of his observations,—

" 1st. That varicose veins of the lower extremities are not more common in cases of ovarian disease than in pregnant women;

2nd. *When it exists, it soon subsides after either tapping or ovariotomy;*

3rd. If the tumour is confined to one side, the corresponding leg is usually affected; but when the tumour is central or bilateral, both legs are equally affected;

4th. When the veins are enormously affected—especially if clots or phlebolithes are found—the tumour is usually fixed low down, below the brim of the pelvis, or does not rise out of it even after tapping."

Dr. Clay, of Manchester, has obliged me by stating that " of some two to three thousand ovarian cases " he " has not noticed varicose disease in more than half a dozen cases; and as most of these were in women who had had many children previously, the varicosity probably existed before the ovarian disease set in." Dr. Clay adds, however, that he frequently sees it in complicated cases (ovarian and uterine combined), and still more frequently in uterine enlargements and in pelvic tumours, where neither uterus nor ovaries are involved. It has, in Dr. Clay's experience, been present also in cases in which some uterine enlargement has blocked up the pelvic passage, which is seldom the case in ovarian affections alone.

These observations, supported by my friend Dr. Murray, but in which other eminent obstetricians do not entirely concur, agree with my own. They are, perhaps, of little moment. The main question as to the *production* of subtegumentary varicosity by these abdominal and pelvic tumours must, I think, be determined in the negative. Still they must embarrass veins already varicose, and by pressure on the systemic veins may give rise to varicosity of the *tegumentary* form.

K

The hypothesis that *utero-gestation* is a cause of varicosity is, for the reasons just alleged in cases of ovarian and uterine tumours, equally untenable. Whatever be the condition of the crural veins prior to utero-gestation, they usually relapse into it again at its termination, although, perhaps, not without some slight aggravation of any pre-existing dilatation.

In thirty-five consecutive cases which came under my own notice, in which utero-gestation and crural varicosity co-existed, the tegumentary as well as the subcutaneous veins were varicose. The ages varied from twenty-three to thirty-three, and although the attention of the patients severally was drawn to the disorder during pregnancy—either at the second or third month, from more or less discomfort in the vessels—yet in every case in which the circumstances were clearly remembered, the varicosity was said to have existed prior to its occurrence, and in most instances to have been connected with some disturbance of the catamenial functions.

It is said by Chaussier that varicose veins sometimes supply a kind of reservoir for unused blood during utero-gestation; and that therefore the act of bandaging them at this period is dangerous, inasmuch as it may lead, as in cases related by him, to abortion.

Pressure on the *iliac vein* either by an impacted colon, or on the left by its associate artery, may impede the free return of the blood from the parts below. In the course of my dissections, I found striking illustrations of this fact. In one body, the iliac vein, on the distal side of the artery, was permanently dilated by the pressure of the artery upon it to double its normal size; the dilatation was limited to this segment of the vessel. In another, with disease of the aortic valves, the systemic veins were dilated above the artery to the heart. In neither was there varicosity or dilatation of the crural or any other veins.

Mr. Kingdon and Mr. Langdon have favoured me with an exceedingly valuable abstract of cases of hernia, complicated with varicosity, taken from the Report Book of the City of London Truss Society. It shows that the number of such cases is exceedingly small; that varicosity is as common with inguinal as femoral rupture; and that in nearly half the cases it is on the side opposite to the hernia. I regret I cannot give the abstract entire.

If complete obliteration of some large pelvic or abdominal trunk fails of producing varicosity in the lower limb, it would scarcely à *priori* be expected that *obstructive cardiac* or *pulmonary* disease could have that effect, and for this obvious reason;— in venous repletion from central obstruction, collateral compensation *is* not possible, whilst it *is* possible when its cause exists in most eccentric portions of the venous system. There is, besides, a want of correspondence between the periods of life at which these disorders respectively begin and culminate. But supposing varicosity to exist, and to be succeeded by obstruction to the systemic trunks from causes such as those under consideration, would the varicose veins be in any way influenced thereby? Barry and Poiseuille conceived that the influence of respiration is *favourable* than otherwise to the *aspiration* of the blood; or, in other words, that the reflux of the blood from the heart during expiration is more than counterbalanced by its stronger influx during inspiration; and that therefore the thoracic functions aid the return of the blood even from the remotest parts of the system. Berard alleged, on the contrary, that it is impossible to *aspire* a fluid through a membranous tube—a view which the use of his manometer contributed much to support. Hence, with the exception of certain veins, the mouths of which, from their anatomical connections, are prevented from collapsing—such as the jugulars, the vena cava to the diaphragm, and the veins of the portal system—the flow of the blood in *healthy* extra-thoracic veins is not materially influenced by thoracic aspiration. But the case is altered should any of these veins become dilated and otherwise diseased, for, with a certain concurrence of morbid changes, of which dilatation is one, veins, before independent of thoracic movement, are thereby brought under its influence. Every one is aware of the impulse communicated to a saphenous varix by the act of coughing; but do thoracic impulses extend beyond the spot where they are detectable by the finger? Dr. Anstie was so good as to aid me in quest of such results by employing the sphygmograph in several cases of varicose disease. In one case, that of a woman with cardiac disease and severe crural varicosity, this instrument detected a wave in a large vein on the inner side of the calf; but it was inspiratory. In a second case, that of a man, æt. 38, otherwise healthy, the sphygmograph detected a wave in a varicose vein in a similar situation; but the wave was expiratory, and in unison with the action of the abdominal muscles. In some cases of uncomplicated varicosity an expiratory wave was detected in the veins; whilst again, in others, in which the conditions appeared to be very nearly alike, no such wave was indicated.

It is difficult to explain the apparent anomaly; it certainly is not dependent upon defective valvular endowment; and time was not given me to follow the subject further. It is, however, manifest that through certain other changes which take place in the walls of the veins that are varicose, the current within them is brought beneath the influence of the thoracic movements, inspiratory as well as expiratory. The latter, however, appear to be the most constant as well as powerful agents in determining a wave; and hence the danger of such movements as coughing, straining, or lifting in cases of aggravated varicosity.

The following cases may serve to illustrate what I have just advanced:—An old lady, æt. 65, came to the Great Northern Hospital. She was bulky; her face was remarkably livid from venous congestion, and her breathing difficult, and aggravated by exertion. She had an ulcer on the leg, near the ankle, surrounded by an extensive network of severely varicose veins, which, she said, constantly became distended almost to bursting whenever her breathing was more than ordinarily laborious. On watching the case I became convinced that her statement with regard to the correspondence between the venous distention and the dyspnœa was correct. I divided the veins extensively by incisions on either side of the ulcer; the bleeding, which was for the moment violent, was easily suppressed by bandages and posture; the dyspnœa was much relieved; the ulcer healed.

I saw a similar case recently in a lady who had crural varicosity with diseased heart—a loud bruit, with irregular and tumultuous action. Her breathing became suddenly embarrassed; legs œdematous and dusky, and the varicose veins large and painful. She was treated for her chest affection, and the leg symptoms subsided concurrently with the dyspnœa.

Chaussier relates a case of a lady affected with asthma, crural varicosity, and œdema. Compression of the limbs by a bandage invariably aggravated her dyspnœa, which lasted until the bandage was removed. She nevertheless persisted in the use of the bandage; the dyspnœa increased, and she died of hydro-thorax.

The series of agencies just reviewed cannot be said to be *causes* of *subtegumentary* varicosity. They can only have the effect of somewhat aggravating existing disease. But they appear to conduce to its development in the *tegumentary* radicles.

The *third* class includes but one cause, viz., *catamenial* disorder. Frankenau * cites cases, and amongst them the following, to show that the diseases of menstruation constantly contribute to the development of varicose disease:—A young woman having ceased to menstruate for two years, had an attack of severe illness. The veins of her leg became varicose, bled spontaneously, and she recovered. Briquet mentions several cases of like import. A case was under his observation in the Salpétrière of a woman, æt. 53, who had varicosity of the left leg. It appeared at the age of 15 or 16, during a period of difficult menstruation, and was ultimately followed by an arrest of this function. At each subsequent period a varicose vein bled, sometimes violently, for four or five days; a slight oozing followed for a day or two, and the vein healed. This vicarious hæmorrhage continued for six years, but on each occasion from a fresh opening in the vein, so that, at the end of that period, the entire vein was covered by cicatrices.

The following case I saw with my friend Dr. Rayner:—A young woman, æt. 28, florid, but somewhat spare, suffered from varicosity of the left leg and thigh, which began coincidently with some catamenial difficulty ten years ago. The

* Miscellanea Academiæ Curiosorum Naturæ (1684).

right limb was slightly varicose as well. Although she menstruated regularly, she was periodically affected with severe headache, which subsided on the occurrence of spontaneous hæmorrhage from these veins. During the last three years the veins have lessened in size, and at the same time there has been a gradual diminution of the menstrual flow. The headaches have, however, become concurrently more and more severe, and the patient is now maniacal. I might relate other cases bearing similar testimony to a relationship between catamenial and varicose disorders.

Of thirty-four cases of crural varicosity which I investigated in succession, the disease was clearly associated with such disturbance in fourteen. In one of these the catamenial discharge was profuse; in three it was altogether suppressed; in another it had been profuse for the three years preceding the appearance of the varicosity, but had subsided; in six it had been scanty and painful; and in the last the veins of the leg were distended and painful at each period, although the function itself had been healthily performed.

The varicosity in most of these cases was, however, of the mixed type — tegumentary and subtegumentary; and we therefore naturally turn for its production, in accordance with our previous conclusions, to a combination of physical with functional agency; muscular activity in conjunction with some embarrassment of the deep venous system through catamenial disorder. But I believe it will generally be found that, with catamenial embarrassment alone, the varicosity invariably assumes the tegumentary form.

We now come briefly to consider the *fourth* class of influences, — the *predisposing;* on which I must be very brief.

Heritage appears, as Mr. Callender observes, to be an efficient member of this family of causes. I have known remarkable instances in which the tendency has apparently been transmitted from one generation to another. In one, several members of the same family, who had a varicose parent, were afflicted with varicosity; and in another, the grandfather and the father died of hæmorrhage from varicose veins, and the next descendant had reason to fear a similar fate.

Clinical observation abundantly proves that exhausting diseases, whether affecting the coats of the veins themselves, as fatty degeneration, or the general system, as, for instance, obesity, fever, rheumatism, gout, &c.—predispose to varicosity. In two cases which came under my notice, varicosity was preceded by acute rheumatism of a severe kind,—the one a man æt. 35, the other a woman æt. 38,—the disorder was bilateral; in the former, the left leg was affected most severely; in the latter, the right.

Mr. Paget* has described a form of gouty phlebitis, and relates an instance in which it was followed by varicosity. Professor Laurie, of Glasgow, in a letter in the same journal, mentions his having himself had an attack of typhus fever, which was followed successively by phlegmasia dolens and crural varicosity. In some of these cases the muscles of the limb are said to undergo enlargement.

Lastly, *injuries* to the limb—fractures, laceration, and burns—are not unfrequently followed by varicosity. Of these causes fractures are by far the most common. Their effects extend to the neighbouring venous trunks, and cause obstruction, either

* Reports of St. Bartholomew's Hospital.

by direct pressure, or by exciting local phlebitis and thrombosis. The varicose veins that follow span over the interspaces within which the injuries occur. In a case of broken thigh high up, the veins between the saphena below the seat of fracture and the superficial pubic and epigastric veins became varicose; whilst in a case of broken tibia the dorsal veins of the leg became similarly affected.

The *etiology* of *subtegumentary* varicosity may be thus summed up. Its immediate cause is *obstruction* of trunk veins through saphenic repletion, most usually through excess of muscular exercise, or direct injury. It is favoured, however, by a number of predisposing agencies — heritage, idiosyncrasy, constitutional debility, acquired weakness in the vein coats from fever, gout or rheumatism, degeneration of their tissues, and it is susceptible of aggravation by any impediment to the blood in its course through the large systemic veins, the lungs, or the heart. With degenerate vein walls the varicosity is prone to extend itself through branches to their extremest ramifications, which is not the case in simple varicosity, or varicosity without any such previous or concurrent change.

Tegumentary varicosity is induced by obstructive repletion of the systemic veins, through the pressure of abdominal or pelvic tumours, probably by that of the left iliac artery, or of an over-weighted colon, pulmonary or regurgitant cardiac disease, catamenial disturbances; and is eminently favoured by obesity and general want of tonicity in the fibres of the dermis. It may always be regarded as symptomatic of systemic venous obstruction; and will generally serve as a guide to its seat.

I come now to speak of the *treatment* of varicose disease.

After many centuries, during which the *cure* of varicosity has been attempted by various means, some with the view of effecting the removal or obliteration of the veins by processes which would have done credit to the genuises of the Holy Inquisition, and, of course, with varying amounts of that inevitable success which is said to attend all new remedies, a reactionary feeling has set in, and with it the suggestion of a reasonable doubt whether the condition of the patient has in the end been really benefited by these proceedings; or whether, whatever has accrued from their adoption, in the way of benefit, ought not rather to have been accredited to the rest and other simply palliative measures with which their employment was accompanied.

M. Verneuil regards the idea of a "cure" of varicosity as "une abstraction faite des moyens chirurguaux :" and Mr. Syme, in a private communication, kindly informs me that he "regards with entire distrust all the means that have been proposed for effecting a radical cure of varicose veins, deeming it fortunate that so much good can be done in the way of palliation." On the other hand, modern works continue to teem with instances in which the credit of curing the most advanced stages of varicosity is assigned to the needle, ligature, caustic, or some still more heroic remedy.

But what are we to understand by "curing" varicosity, or, if you will, a varicose vein? Is such a vein curable? The answer "No" has already been given. Such a vein is incurable. Does it mean that state of the limb and its veins which often follows an attack of phlebitis? I believe such "cures" are generally either hypothecated or misconstrued. If permanent benefit has accrued to such a limb after, and it often does follow, such an attack, it is, for the most part, due to the rest and other means employed, and not to the phlebitis, for the state of the veins is not found to be materially altered when the attendant local mischief shall have cleared away. The limb is better for a time, but the friendly stocking cannot be cast off, nor the pristine exercises again indulged in. But occasionally there is, after such an attack, a very near approach to a "cure;" for, as I have said, a varicose vein sometimes becomes plugged up wholly or partially, and the circulation is in great part restored to its normal channels. Nevertheless, the patient is advised by his own consciousness as well as by his surgeon that he cannot return to his former active habits without peril. This kind of cure, however, takes place generally by a slower and somewhat different process.

For many a century a varicose vein was looked upon much as the divines of the sixteenth, as humorously suggested by Washington Irving, regarded the natives of the New World. "They had," argued these worthies, "nothing of the reasonable animal but the mask; they were of a hideous copper complexion, and that was the same as though they were negroes; and negroes are black, and black is the colour of the devil, therefore they ought to be exterminated." At a late Roman period even they were doomed on account of their interference with personal beauty; a pair of unblemished calves appear to have been as much a matter of importance to Caius the Consul as to the modern "Jeames" or "Bumble." Plutarch relates of C. Marius, that "both his legs were full of varicose veins; and that, being troubled at the deformity, he determined to place himself in the hands of his surgeon. Refusing to be bound, he stretched out one of his legs to the knife, and, without motion or groan, bore the inexpressible pain of the operation." But, as we are further told, "when the surgeon was about to begin with the other leg, he would not suffer him, saying, "I see the cure is not worth the pain." Vanity succumbed.

The earliest recorded methods of treatment include puncture, longitudinal and transverse incision of the diseased vein with or without removal of the clot, excision, cauterization, and ligature.

Hippocrates punctured these veins, but did not advise cutting into them, for fear of thrombus. Celsus advised that "omnis vena quæ noxia est, aut adusta labescat aut manu exceditur." In the 15th century Ætius treated incipient varicosity by the topical application of bitumen, pitch, or turpentine; and its more advanced states by cauterization or ligature. Paulus Ægineta, Albucasis, and Avicenna adopted the same treatment. Severinus and Fallopius tied the vein in two places, and excised the intermediate portion. Parè recommended bleeding from the vein, with rest and other appropriate remedies; but, with ulcer, ligature or excision. Guillemeau proposed puncture of the vein in two or three places, and the subsequent application of caustics or the actual cautery; whilst Fabricius suggested the modern plan of dividing the vein between two ligatures.

These modes of effecting what is called the "radical cure" of varicosity include in principle as well as in practice almost all that are in vogue at the present day; and if called upon to decide between them, we should say that, except, perhaps, in manipulative detail, the difference is as that "between tweedle-dum and tweedle-dee."

In later times, Sir Everard Home suggested *deligation* of the *saphenous trunk* on the inside of the knee for the cure of varicosity. The practice, however, fell into disuse through the mishaps which attended it; and it is not likely to be revived, although, according to the *Medical Record* for June, 1867, we are informed that the deligation of large veins is not so serious a matter as Sir E. Home appears to have found it. Dr. Greene, the able Professor of Surgery in the Berkshire Medical College, U.S., has, it is said, ligatured the jugular four times, the femoral six, and other large veins repeatedly, without any of those grave results which he had been taught to apprehend. In the experience of Sir Everard, Sir B. Brodie, Sir A. Cooper, Briquet, Lawrence, and Warren in America, cases of deligated saphena proved fatal; and in others, one in my own practice, the ligature has been followed by results sufficiently serious to lead to its discontinuation.

In 1817, Sir B. Brodie * recommended *subcutaneous division* of the diseased vein itself. The wound generally healed in the course of a few days. Severe phlebitis, however, sometimes followed; and it appeared, says Sir Benjamin, as "though the veins generally healed without becoming obliterated; and in case the cluster became obliterated, others took its place, and no benefit ensued."

In the experience of Ferrall, Carmichael, and Beclard suppuration followed in several instances; and, in that of other surgeons, even death. At length Sir Benjamin, with his usual candour, told Mr. Cooper that he thought the advantage, claimed for this mode of treatment, was due to the rest that it necessitated.

Graefe recommended incisions, two inches in length, in the

* Med.-Ch. Trans.

most knotty parts of the veins, plugging the vessel with charpie, and compressing it with a roller, repeating them, if the disease extended throughout the limb, at the knee and in the thigh.* This plan was approved by Chelius. Briquet recommended excision of the diseased veins; but only, as far as I can learn, when they assumed the form of varicose tumours.

Mr. Colles, of Dublin, devised a truss with a spring pad, wherewith to compress the saphena in the groin.

Mr. Hodgson wisely fell back, in accordance with Sir B. Brodie's views, upon rest, the horizontal posture, and compress as the best remedies for varicosity. M. Nelaton also advises compression of the superficial veins by a stocking or bandage; but in a manner so as to resemble the pressure of the deep fascial aponeurosis upon its subjacent structures, with the view of preventing the establishment of a collateral circulation by forcing the blood from the superficial into the deep veins. M. Verneuil proposes the same end by like means. In some instances M. Verneuil has been satisfied with the results, but in one it induced œdema with repletion of the varicose veins about the heel and the region of the tendo-achillis, which was supposed to have arisen from the failure of the compress to command those deep veins from which the varicose channels were supplied. Is it not more probable that some of the deep trunk veins were obstructed? Mr. Startin supposed that he could supply the loss of hypothetical valves in varicose veins by a spiral bandage, a principle of treatment which Sir B. Brodie seems to have been the first to suggest, and in accordance with which he recommended that strips of adhesive plaster should be tightly drawn across the

* Graefe's edition of "Bell's Surgery."

vessels. As I have shown, these veins have no valves, and if they had, the valves would not, as the bandage inevitably would, offer any resistance to the current.

Mr. Nunn has proposed elastic stockings and rest for cases in which the blood is hindered in its return to the heart; and, in other cases in which the varicosity is supposed to be caused by columnar pressure, to force the blood into the deep veins by the adjustment of a series of pads along the diseased vessels, in the sites of their greatest dilatation.

These practices were followed and, to a certain extent, have been supplanted by the simple ligature of the varicose vein.

Various modes of applying the ligature, and ligatures of different textures, have been alike suggested. Minkiewiez[*] has discussed the relative value of metal and silk in a paper occupying nearly seventy pages. Davat, of Aix, after experimenting on animals, is said to have been the first to suggest a needle beneath the vein, crossed by a ligature in the shape of the figure 8. This plan has had the commendation of Velpeau and Begin, and is now, I believe, usually adopted by Sir. W. Fergusson. I have myself employed it largely, and do not know of any better contrivance, when the ligature is required.

Mr. Wood [†] uses two needles, which are so adjusted as to make complete, equable, and constant lateral as well as linear pressure on the vein, without including the skin in the compress. It is said to obliterate the vessel, and at the same time to prevent any danger from the inbibition of putrid matter or pus.

Mr. Lee has recently proposed a modification, which is, per-

---

[*] "Archiv fur Pathologische Anatomi," vol. xxv., p. 193, 1862.
[†] Medical Times and Gazette, 1860.

haps, as free from danger as any plan that has preceded it, if not more so, and as effective. Two pins are placed under the vein at the distance of three-fourths of an inch from each other, the vein being pressed against the pin by india-rubber ligatures, and divided between them subcutaneously. Mr. I. B. Brown tells me he uses simply a piece of fine silver wire, with which he surrounds the vein.

The next method of obliterating veins to which I shall refer, is by the *formation of a thrombus or clot* within the vessel, without ligature.

In this country Sir B. Brodie was, I believe, amongst the first to introduce this method. He applied potassa fusa along the track of the vein, wherewith he penetrated the skin to the vein coat; but Sir Benjamin was not encouraged by his experience to pursue or recommend it.

Mr. Mayo[*] then employed nitrate of silver for the same object, and reported that in some cases it produced no effect on the vein, whilst in the larger number the vein became hard and appeared to be obliterated at the part where the eschar was made. He had not known any bad results from this practice.

M. Bonnet, of Lyons, approved this plan; but M. Laugier and M. Berard substituted Vienna paste, the one with, the other without incision of the skin. In this country this caustic plan has received the unqualified approval of Mr. Skey, but without the preliminary incision. In the *Lancet*[†] Mr. Skey reports successful results in twenty-five cases, and says that the plan is free from comparative danger. Mr. Skey makes as many as from ten to eighteen eschars along the course

---

[*] "Outlines of Human Pathology."   [†] October 5th, 1861.

of the vein, according to the extent of the disease. The paste he uses is composed of powdered lime and caustic potash, in the proportion of three parts of the former to two of the latter, moistened with alcohol. He selects the most prominent parts of the vein for the eschar, and says that an eschar of the "smallest size is sufficient at once to obliterate it."

M. Nelaton has both seen this treatment adopted and tried it himself; but, as he says, with very little advantage—at all events, not more than may be attained by a simpler and less dangerous procedure.

More recently, surgeons have been somewhat startled by the proposal to obliterate varicose veins by injecting diluted tincture of the sesquichloride of iron into them. It is said to have originated with Dr. Pravaz, of Prague. In the "Archiv für Pathologische Anatomi" there is an elaborate article on this subject by Dr. Ellinger, who claims the credit, whatever that may be, of introducing the practice into Germany, from whence it had been "excluded by the phantom of embolism." In this paper the author alludes to four cases:—two of erectile tumour and two of varicose veins—into which the remedy, in the proportion of one of the tincture to thirty of water, was injected. As to the results, we are only told that " the latter are doing well in a short time." In France, as appears from the same paper, the practice has had a much more extensive trial, which has resulted in eliciting from M. Broca and Adolphe Guerin expressions of disapproval on account of its being both dangerous and ineffective.

Dumarquay, Voillemier, and Maissoneuve have, on the contrary, accredited to it the highest praise. Voillemier states that he has used it in a large number of cases. Maissoneuve tried it in hæmorrhoids, varicocele, and varicose veins, and regards the procedure as innocuous. Serious accidents have, however, followed its adoption in some of the Parisian hospitals, as elsewhere; but Herr Ellinger is nothing daunted, upon the common principle that "accidents will happen," or, as he puts it, "death will follow the treatment of other trivial cases."

In this country the method was tried by Sir Henry Thompson in 1864,* and by Mr. Erichsen.† Sir Henry tried it in two cases. In one it was followed by severe pain in the groin in the course of two or three days; and the woman, æt. 44, did "pretty well." The vein was obliterated. In another case it was unsuccessful.

In two or three cases, in which Mr. Erichsen used this injection, it was followed by circumscribed abscesses and sloughing of adjacent parts; but the patients survived. Mr. Erichsen regards the practice as dangerous, and prefers to obliterate the vein by compressing it between a needle and a piece of bougie laid lengthways along the vein. Of two hundred cases thus treated there was not one in which it was followed by phlebitis or pyæmia. Its curative powers are, however, not stated.

Other modes of obliterating the veins have been proposed.

Davat passed a needle once or twice through the vein, and tied the coats of the vessel against it. Frieke substituted a thread for the needle which he made use of as a seton, and reported twenty-five cases of cure by this method.

Sanson used a couple of oval discs, connected by a spring, and closed by a screw,—a plan analogous to the *button* suture of

* Lancet, Oct. 2, p. 351.   † British Medical Journal, p. 141.

Rozeman, of New Orleans; whereas Vidal used a similar contrivance in the form of a clamp spring.

Galvanic puncture has not been without its advocates in this as in most other affections; as well as other methods, to which it is needless to refer.

All these modes of dealing with varicose veins are designed to effect the cure of the disease either, 1st, through the obliteration of the veins themselves or of their trunks,—evidently under the misconception that the *disease* is limited to the vessel or vessels selected for operation, or at all events to these and any varicose branches which may spring from or are connected with them; or, 2nd, by diverting the blood from its new and diseased courses into its former hypothetically normal channels. The *principle* involved in both is *cure* by *obstruction*. Now, if what has been advanced has any foundation in truth, the principle must be faulty, inasmuch as it involves the anomalous proposition to cure obstruction by adding thereto fresh obstruction,—a proposition upon a par with the notion of damming a stream in order to relieve the overflow of its tributaries; and Nature has persisted in thwarting a procedure based on such contradictory premises.

I must, however, except Mr. Herapath's proposal to divide the edges of the foramina through which the saphenæ pass to their trunks, for the cure of varicosity. Mr. Herapath carried it out, in some cases, so far as to divide the falciform edge of the fascia lata, and, as he thought at first, with decidedly beneficial results. In a letter to Mr. Chapman some years afterwards, Mr. Herapath acknowledged that any benefit which accrued at the time from this proceeding was but transient, and that he had lost confidence in it. The principle was correct—the cause of failure obvious.

If, then, varicosity is not to be cured, or perhaps even indirectly relieved, by obliterating or otherwise obstructing the diseased vessels or their trunks, what are we to do? What principles of treatment are we to adopt? I answer, 1st, that so long as varicose veins are capable of aiding in circulating the blood, though with comparatively trifling efficiency, we must (*a*) relieve the general circulation of the limb as far as possible from those causes of embarrassment in which their disease originated; (*b*) preserve the vessels in that state of usefulness to which they may have been reduced, or render them still more useful by giving artificial support to their deteriorated walls; (*c*) remedy any contingent disorder of the vein as far as it can be remedied; and (*d*) adopt such general measures as shall have the effect of indirectly imparting strength to its tissues. And 2nd, (*a*) in the event of any portion of such vein becoming so hopelessly deteriorated that it can no longer aid in furthering the circulation, especially if it be irremediably painful on, or without exercising the limb; or (*b*) if the vein shall have given way, or appears, from attenuation or other conditions, liable to burst without forewarning;—under either of these circumstances the particular segment or entire branch must be obliterated.

In the early stages of the disease the exemption of the limb for a time from muscular exercise, resting it in a recumbent position, the use of the flesh-brush, as recommended by Mr. Vincent, the cold douche, elastic bandage, and such remedies as are calculated to remove any impediment to the flow of the blood through the systemic veins, and to restore to the diseased vessel its physical powers, will tend powerfully, not only to check the further progress of the disease, but in many cases to regain for the vessel, and for the limb too, much of the power and ease of which they had been deprived. But these means must be discreetly persisted in for some time after even such results have been procured, or they will not be retained. Even in advanced stages of varicosity, results that amount to a virtual cure have been, in my experience, arrived at by the persistent use of these several means; so that, after a time, a burnt-out varicose vein, an old stain, and perhaps a cicatrix have been the only remaining evidences of a disease by which the patient has been for a time bedridden.

The *posture* of the leg in the treatment of these cases is of the first importance. The foot should be kept above the level of the pelvis, so that the return of the blood shall be favoured by its own gravitation. This position was, I believe, suggested by Brodie, and has since been recommended by Liston and Hilton.* For this purpose I have recommended, and with benefit, that for the night the bed should be so arranged as to necessitate the sinking of the pelvis; or that a cushion, in the form of a collar, should be placed around the ankle.

Bandages in the form of stockings should be made of tolerably firm but still elastic material, and without *ribs* of any sort. They should also be applied with different aims, according to the nature of each particular case. Sometimes they are borne even applied tightly, at others only with the least possible amount of pressure, whilst again they are intolerable if they compress the veins ever so gently. This unequal toleration of compresses depends upon conditions of the deep veins which can only thus be ascertained. In cases of simple, uncomplicated varicosity there is no reason to

* Lancet, 1855.

suspect any deep venous obstruction; and it is then perfectly feasible and right to attempt to contract the diseased vein with the view of diverting the surplus portion of its current into the deep veins; but not even then should the compress entirely *close* the diseased vein, for it has always been a channel, and an important one too. In other cases where, from the state of the skin, the irksomeness of the compress, or the supervention of œdema, some obstruction of the trunk veins, especially the subaponeurotic, is suspected, the bandage should be applied, rather with a view to the *support* of the vein coats than to the diversion of the stream. The bandage should always be adjusted, in respect to tightness, so as to give a sense of comfort and relief, and should on no account give uneasiness.

It is not necessary that the whole limb should be included within the *compress* in case of limited varicosity. It should encompass that portion only in which the varicose veins are situated. For *advancing* varicosity the bandage should be more widely applied.

In painful states of the vein, from over-distention, often met with during pregnancy, the compress is not well borne; soothing remedies, opium, rest, and posture afford the best relief. In such cases I have punctured the vein, and let out a few ounces of blood, with marked benefit. The use of quinine, steel, zinc, and other tonics is often indicated, with rest and compress, and found to be of great value.

The veins are at times painful, doubtless from chronic disease of their coats through over-distention; a condition for which rest, with elevation of the foot, is also absolutely necessary. Other remedies are, however, required in order to afford more than temporary relief; such as the frequent applications of hot water, or of poppy or hemlock fomentations. If inflamed, a few leeches, followed or not by a blister, *alongside* the vein, or the application of a strong solution of nitrate of silver, or of the salt itself along its track, will afford relief. It is well, however, to avoid creating sores on such a leg. If the patient is feeble, the topical application of opium or belladonna is indicated, with tonics, especially steel, and morphia at night. Above all is it important in all cases to keep the lower bowels cleared out either by aloetic purgatives, or others if aloes be not borne, or by enemata.

I need not occupy your time by detailing the symptoms, issues, and treatment of acute phlebitis when it attacks varicose veins, 'beyond reminding you of Mr. Lee's proposition, derived from a remark of Mr. Hunter, to place a compress on the vein, at the limits of the disease, in the centripetal direction.

But varicosity gives rise to subjective symptoms which an acquaintance with its pathology can alone render an account of, or aid us in our attempts to remedy; such as cramps and other pains referred to the muscles, muscular fatigue, neuralgia, pains and even numbness in the sole of the foot, weakness of the ankle; and, as alleged, to elevation of temperature, and excessive perspiration.

Œdema, and both muscular hypertrophy and atrophy, when met with in connection with varicosity, must be referred to some one of those morbid changes of which I have spoken—intra-muscular dilatation and other disease of the vein coats, thrombosis, &c., and must be treated accordingly. The especial means I need hardly point out. I may just say of cramp that I am inclined to refer it to some difficulty in the intra-muscular venous circulation occurring on the first muscular efforts after a period of rest. It is usually overcome by some one determined effort, as in standing firmly on the affected limb.

For *asthenic* varicosity, in addition to local treatment, general constitutional treatment will have to be adopted, such as tonics, fresh air, exercise to a very moderate extent, generous diet, wine, and the avoidance of all exhausting influences.

The various modes of permanently *obstructing* or *obliterating* varicose veins which I have laid before you are, for the most part, open to the objection that they accomplish their object, either directly through the medium of coagulum, or by other direct processes but still not without its production, within the walls of the vessel.

I need not say that the formation of a thrombus in a vein is on many accounts a procedure fraught with danger, and ought, if possible, to be avoided. It is true, both thrombi and emboli frequently exist in these vessels, and probably without doing any serious mischief; but there is an important distinction to be made between clots formed slowly, *pari passu* with, and in obedience to, changes in the vein coats as well as in the general scheme of circulation, and such as are procured traumatically; viz., that whereas the *former*, the idiopathic thrombi, are often reparative and therefore comparatively innocuous, the latter are dangerous, and have occasionally been followed by mortal consequences. Instances of pneumonia and pulmonary abscess not unfrequently occur in which these affections are due to unsuspected embolism. A gentleman last summer was accidentally shot by a bullet, which lodged in the upper and inner part of the thigh. There was no local mischief of consequence beyond suppuration of the track of the wound and, subsequently, œdema of the leg.

After some weeks pneumonia of a low type followed, with severe bloody expectoration, which almost cost him his life. In another instance, a gentleman in Devonshire injured a varicose vein in mounting his horse. Slight phlebitis followed, and afterwards an attack of pneumonia attended with profuse purulent and bloody expectoration. The attack was repeated after a fortnight's interval of apparent convalescence, and his life was for a time again despaired of. I believe these are by no means rare instances of embolical mischief, nor the worst. In a third case (one of my own) after tying a large varicose vein, death resulted from like pulmonary disorder.

The obliteration of a vein should be effected without the risk of a clot being formed within it. Hippocrates, as we know, was alive to the importance of this precept; and it can be carried out by employing the ligature. The kind of ligature, as well as the precise mode of using it, is of much less moment than the preparation of the vein, and the treatment of the limb at the time and after its application.

The portion of condemned vein should first be carefully examined, both with and without its blood, and the more solid parts chosen for deligation. Ligatures should then be placed beneath the vein at distances of from three-fourths to an inch apart. I have generally used a fine needle of soft metal, crossing the vein with a silken or hempen thread, in the form of a figure of 8. Mr. Brown's loop of silver wire, without the intervention of a needle below the vessel, is perhaps as good a plan. The blood is then to be poured out of the vein by elevating the foot, or, if necessary, by a bandage on the distal side of the vessel (but this is seldom required), and the ligatures are to be tightly tied.

The leg should then be suspended, with the foot above the level of the trunk; the ligatures removed on the fifth or sixth day; and a bandage firmly applied, with an intervening pad of lint, along the course of the obstructed vessel. In the course of ten days the patient may move about, but an elastic stocking must be worn for some time. In this manner a vein, or portion of vein, may be successfully obliterated without risk of thrombus or serious phlebitis.

It is somewhat humiliating, after having led you so far, to be obliged to confess that the end of our inquiry, as far as it is calculated to indicate remedial measures, is rather to discourage than to foster our trust in them. Still it cannot, if based upon facts, have been otherwise than worth pursuing, if it recall us from speculative efforts to *cure* varicose disease to a matter-of-fact calculation of what may be attempted with reasonable expectation of success, and what cannot without the probability, amounting almost to a certainty, of failure. We learn the measure of our powers, and should be instigated to renewed efforts to improve them on a scientific basis, which may yet, I believe, be largely and profitably extended.

The skin of the lower part of the leg, most usually between the ankle and the calf, is often discoloured. The *colour* is that of *bronze*, mottled with lighter shades of the same tint, and toned off into that of the healthy skin. The stain is often permanent, for on the healing of an ulcer in such a portion of skin, the cicatrix is of the same colour, although of a lighter shade. Sometimes it is dark and polished, like the skin of the negro; at others it is coarse, even to ichthyosis.

The skin thus stained differs in tension and density in different subjects. Sometimes the stain constitutes the only visible change, when it usually occurs in comparatively small, abruptly defined, and irregularly shaped patches, and marks the sites of old syphilitic eruption or ulcer. In other cases it overlies portions of integument that are either soft or boggy, or more or less elastic; whilst in a third variety the integument is more or less hard and unyielding,—so much so as in these respects, in some cases, to resemble india-rubber. In this form the discoloration extends to the entire circumference of the limb, and is accompanied with marked contraction in its size, as well as thickening, which pervades the skin tissues to the deep fascia, and often to the fascia itself.

In the uncomplicated form, the stain appears to be due to some change in the *rete mucosum*.

In the second variety—the "tissue fungueux" of Briquet—it is due, in part to the occupation of the skin by an intricate network of veins, tortuous and dilated to their capillary outposts, in which the vascular elements predominate at the expense of the dermoid, and in part to the discoloration of the inner coats of the diseased vessels. In one case (Dis. 2) the internal saphena and its branches were enormously dilated, and there had been long-standing embarrassment of the stream through the right side of the heart. In a severe case of tegumentary varicosity (by way of contrast) (Dis. 1), there was intense congestion of the *internal* saphenous, a dilated posterior tibial, the popliteal and femoral veins. The *external* saphena was incompetent.

In the third, in which the relation of the cutaneous and vascular elements are, as compared with the skin in the second variety, reversed, the stain is due almost entirely to the effusion of blood in the upper lamina of the dermis. It may extend below these lamina but only in limited patches. It is then generally in connection with disorganization of some of the smaller veins. The blood either softens into a kind of pulp, and thus becomes blended with the tissues, or it may remain, without softening, as infinitely minute clot-rolls—veno-capillary thrombi—histologically in less intimate, but not less permanent union with them.

The conditions under which this blood extravasation takes place differ, but I believe it is generally the result, either of obstruction in an important branch through inflammation or clot, or of a chronic form of "hæmorrhagic capillary phlebitis." Cruveilhier describes the acute form of this disease. The skin is discoloured to such a degree as in all respects to resemble mortification; the subcutaneous tissue and the small venous ramifications are distended with black blood; the saphenous trunks inflamed and stretched like cords, as well as their subcutaneous branches and deep trunk veins to the external iliac, but not the collateral branches; and the muscles are pale. Unlike ordinary purpura, in the bronzed skin the blood and the vessel appear alike to perish. It is not often that opportunities occur of watching the development of these stains. A professional friend had an abscess in the deep parts of the foot, followed by inflammation of the long dorsal vein, and that portion of the saphena to which it was tributary; the foot became œdematous, and a carbuncle which formed on the site of the minute ramifications of this vessel, on healing, left the skin permanently hardened and "bronzed."

The series of changes which takes place concurrently in other parts of the tegumentary layer and its connections begins in the fat layer. At first the fat in the lobules becomes coarse and granular; it then gradually disappears, whilst the areolar elements which form the network thicken in the same ratio, and ultimately occupy its place, as a kind of coarse fibrous tissue, the fibres of which arrange themselves into parallel bundles. If this tissue be cut through, minute points can be seen with the naked eye at somewhat regular intervals between the fibres. These are the open mouths of what were originally the veins of the fat lobules, out of which the blood can be expressed, but does not flow spontaneously.* When the whole of the fat layer has thus degenerated, it seems to communicate its action to the contiguous structures, superficial and deep. These thicken, and form, with it, in the end one dense and almost indivisible tegumentary layer. When the fat is thus removed from the integument, it sometimes becomes deposited in thick layers below the fascia and amongst the muscles. As these changes proceed, so the vessels, as they come within their sphere, become involved in them. I cannot speak with certainty of the arteries, but the veins undergo a process by which their inner coats soften, and ultimately disengage themselves in great measure from the areolar sheath, which, on the other hand, coalesces with the dermoid tissue, so that in advanced cases the channel, seriously lessened in capacity, seems to have been bored or wormed out of it.

The saphenous trunks, however, are not so early involved as their branches; but as these changes proceed, so they become liable to suffer compression, wasting of their walls—which assume a

* a, b, c, fig. 5, Pl. IV., is intended to give an idea of these several stages in the process of degeneration here referred to. The fibres in c should have been represented coarser.

pinkish hue,—and ultimate obliteration, with or without the formation of a clot. The bone, as Mr. Hodgson observed, will sometimes, in the case of the external, thicken and form a groove for its protection, when thus exposed to the risk of obliteration. The nerves, too, suffer compression.

In one case of skin discoloration with induration both saphenous trunks were obstructed, and there were strange thrombi in the deep veins. In four cases the external saphena was obstructed by clot. In one of these this was the only morbid condition; in the others there were complications, as follows :—In one the deep veins to the triceps foramen were dilated and gorged, and both tibial arteries were encrusted; in another, one of the posterior tibial veins was obstructed by clot; and in the third the same vein had become obliterated, whilst the accompanying vein and the anterior tibial veins were dilated, and their associate arteries encrusted. In two other cases the obstruction was confined to the deep trunk veins; but in each the arteries were considerably diseased. In one, both anterior and posterior tibial veins were thrombose, whilst in the other, a posterior tibial vein was obstructed, and its associate dilated.

We have, then, but to link the morbid conditions constituting, and associated with, this affection of the skin to those several conditions of the vascular system just described, and with which they have been found to co-exist, and it is difficult to avoid the conclusion that, in the cases related at all events, they stood to each other, in each case respectively, in the relation of cause and effect.

And we have only to admit so much, and the morbid phenomena—viz., fibrous degeneration in the *deeper*, and veno-capillary congestion with thrombosis in the *superficial* textures, of the skin—admit of ready explanation on the hypothesis of depraved nutrition from venous embarrassment.

There is, however, an induration without any but a faint pink discoloration, which consists in very much the same histological changes in the integument, but which, as we have also seen, is connected with arterial disease, and is more frequently associated with ulcer than the induration with bronzing. I am not aware of the state of the smaller arterial ramifications in the latter form of induration; but in the former they are very abundant, and form an extensive network throughout the diseased tissues. The venocapillary system shows no evidence of disturbance. But I must proceed to the consideration of ulcer.

Ulcer has been met with—1st, without other *external* disorder; 2nd, with varicosity; 3rd, with discoloured or "bronzed" skin, with or with induration; and 4th, with indurated skin of nearly its natural hue, with, in the *deeper structures*, 1st, venous incapacity, usually from clot obstruction; and 2nd, arterial incapacity from atheromatous or earthy incrustation. Now it requires that we should refer for a moment to the pathology of ulceration, in order more clearly to comprehend its relation to those causes which are of perhaps still more practical import, viz., the *remote*.

An ulcer implies the death and removal of a portion of tissue, of which the surrounding tissues form the residuum. It has been variously denominated ulcer, cancrum, cancer, chronic phagedæna—terms which signify eating away, as though the removed parts had been *eaten* away by those remaining; each consecutive lamella of tissue, after preying upon the one above, having been called upon to yield in its turn as prey to the one below. This is obviously an erroneous view of the process.

The more correct view is that advanced by Mr. Simon, viz., that ulceration is the death and eviction of the lost tissues,—a species of molecular gangrene, with which the absorbent system has no direct concern; the difference between chronic ulcer and acute phagedæna, as far as the mere process of destruction is concerned, consisting in the relative rapidity with which it is accomplished. Very large ulcers are formed by sloughing in very few days; and there is often nothing by which these can be distinguished from others of many months' formation but the difference in the collateral element—*time*. The one ulcer, like the other, after attaining a certain size, usually remains almost stationary. In ulceration, the tissues die, and their death is due to a failure of nutrition, through some vice *whereby the capillary department of the vascular system* fails in the performance of its functions, in consequence of either venous or arterial embarrassment.

Ulcers of the legs have been found to assume every degree of intractability, or rather, to vary in their degree of amenability to treatment, from a comparatively ready submission to the influence of rest, to absolute incurability by any known means. These varied *dispositions* have induced writers to classify them according to some particular conditions or associated disorders to which they have been supposed to owe in this respect their special peculiarity; as, for instance, the inflamed, irritable, varicose, hæmorrhagic, spongy, rodent, phagedænic, "the ulcer with, and the ulcer without action," and the specific ulcer. Now such a basis of classification explains nothing, and informs us of nothing beyond what is revealed to the eye of the surgeon or to the consciousness of the patient. They may tell of accessories by which the process of ulceration once induced is to a certain extent affected or modified; but of themselves they afford very little if any clue to the real character of the ulcer, as this is elucidated by its pathology.

Moreover, experience instructs us that the treatment which appears readily and effectually to cure *one* ulcer, will induce the cicatrization of *another*, but often without permanency of result; whilst upon a *third* it will be followed by a scarcely appreciable, and that but very ephemeral, benefit: whilst all ulcers are, to a certain extent, amenable to courses of treatment adapted to those subsidiary affections by which some are more or less, and others are wholly modified.

The explanation seems to be afforded by the pathological facts to which I have adverted, and the inferences to which they naturally lead; viz., that ulceration, complicated with venous embarrassment, is directly related to the nutritive function, as this is affected by induced morbid conditions of the capillary system:—first, with arterial competency, and second, with arterial incompetency; venous congestion being, in all probability, its immediate and essential antecedent. Hence I distinguish three varieties of ulcer—the *simple, venous,* and *arterial.*

The type of the first variety—the ulcer that is free from all morbid complications, independently of such as are in operation within the confines of the ulcer tissues—is, perhaps, the "broken shin." There are few surgeons who have not had their ingenuity puzzled and their temper tried by the rebelliousness of this somewhat uninteresting surgical trifle, and who have not found it to persist in keeping open until the venous currents of the limb are facilitated by resort to the leg rest. Although it is a traumatic ulcer, it does not essentially differ from that other ulcer which began in pustule, the evidence and result of some constitutional injury, excepting that the latter may be still more indisposed to be cured. Both are equally related in the character of direct effect to that exceptional, but still normal, condition of the venous circulation in the lower limb to which I have adverted ; but made somewhat to vary by the differing accidents of their origination, and other agencies of a secondary character, such as inflammation, certain depraved states of constitution, struma, syphilis, &c., &c., in obedience to which such ulcers will assume a corresponding variety of forms, but be essentially nothing more than simple ulcers still.

And such ulcers *will* form upon legs *with* or *without* varicose complication, and, in either case, equally retain their characteristics, with, perhaps, the exception of being a little more or less prone to venous congestion. The act of *ulceration* in such cases, if properly construed, may be said to be the *way* of *healing* under certain difficulties rather than the way of *ulceration;* for the ulcer will yield to rest and other appropriate local and constitutional remedies, and this equally on a varicose as on an otherwise healthy limb.

But I think it must be conceded that the histological elements immediately concerned in the ulcerative process may, at some undefined stage in the career of an ulcer, become possessed of a certain specific vice, by means of which that process is maintained, although the cause in which it originated shall have ceased to exist—a *habit* (to adopt a happy expression of Mr. Hunter) of ulceration communicated in all probability by a molecule, that has become dead and effete, to that which supplants it. Thus certain ulcers known, for the most part, by their being superficial, but often wide-spread, and having a sensibly thickened base, will perpetuate themselves in despite of almost every variety of *general* treatment, but will yield at once to the *topical* application of some agent capable of destroying the vitiated surface tissues. Such ulcers are usually of constitutional origin; but have survived their cause, and become self-existent. The *simple* ulcer, then, is *the ulcer without complication—except of a purely constitutional character—beyond the limits of the issues immediately concerned in the act of ulceration.*

I now come to the ulcer of the *second* class, the type of which I may describe as an ulcer in the midst of a portion of dense and often discoloured skin; it may be fringed with varicose veins at its lower margin, with *one* or *more* escaping from beneath the upper. The veins in such a case are advanced in degeneracy as they approach the confines of the ulcer; whilst they sometimes become a part of its surface, and share the fate of its tissues. This is apparently the ulcer with "excess of action" of Mr. Syme. What is the relation of the ulcer to the varicose veins in such a case as this? The common response would be that it is essentially that of effect as well as sequence. I must, however, dissent from such a conclusion, and I base that dissent upon pathological facts and reasonings to which I have already invited your attention,* and which connect an ulcer of the kind referred to, albeit with varicosity, with quite other pathogenetical sources—viz., various forms of trunk venous embarrassment.

* See pp. 96, 97, &c.

But how are we to explain the fact that the portions of veins in the most advanced state of varicosity are found constantly in close juxtaposition with such an ulcer? A few facts will, I think, suffice to show that the veins are thus injuriously affected by the ulcer and its processes, and not the ulcer, as is ordinarily supposed, by the veins (although I am far from denying that at this stage of the respective diseases they do exert a mutually deleterious influence upon each other); and for this reason, that the condition of the vein is always in exact accordance with the particular condition of the tissues that surround it. From notes I have made, I find the following instances :—

1. An ulcer in the interspace between two veins, but encroaching on both. Portions of these veins, not elsewhere diseased, but diseased in *contiguity with the ulcer surface:* one softened and filled with clot-pulp; the other in part destroyed, its mouths open on the ulcer.

2. A vein healthy in other parts; but *beneath an ulcer* its coats softened, and its cavity filled with clot.

3. A varicose vein running from beneath an ulcer, and surrounded by discoloured and indurated tissue. The outer coat of the vein consolidated with this tissue, and in the act of disengaging itself from the inner coats, which had softened into a kind of fragmentary tissue, and become incorporated with clot. A segment of the vein above the hardened tissue much more attenuated than the rest of the vessel.

4. A varicose vein beneath an ulcer with a margin of indurated skin. As in the last instance, the vein just above that margin excessively attenuated; below it, contracted; beneath the ulcer itself the vessel completely disorganized.

I think these observations accord with what an attentive examination of these conjoint diseases would daily attest, and go to substantiate the proposition that, in the main, the vein is injuriously affected by the ulcer, and not the ulcer by the vein.

But let me appeal to the experience of those who, adopting the theory of the "varicose vein," have naturally looked to the destruction of the vein as the inevitable remedy for the ulcer.

There are few of us who cannot remember, in the course of practice, how often the problem, "given an ulcer and a varicose vein," has been solved by the postulate, "the ligature;" and with what gratification the cicatrization of the ulcer has been very constantly hailed as its seemingly categorical demonstration. But there are few of us, I think, who, having had opportunities of following such cases into after-life, have not had reason for thinking that, in most, the advantage was short-lived and treacherous; the ulcer and varicose veins having both reappeared. Are we satisfied that even the temporary benefit was due to the ligature? I confess I am much more disposed to attribute it to the rest, and the other means that were simultaneously employed; and amongst them, perhaps, not in small degree to the counter-irritation set up by the operation on the vein; for, as Sir Charles Bell tells us in his "Institutes," "ulcers of the leg are to be cured by a blister applied on the inner side of the thigh just above the knee."

I have now spoken of two classes of ulcers—first, the simple; and secondly, the ulcer constantly affiliated to varicose veins, but which, as I have endeavoured to show, is due to deep venous embarrassment; and which I propose, on that account, to call the *venous* ulcer.

There is yet another species of ulcer to which I must allude, also at times associated with varicosity, but which has, in its worst forms, escaped the designation "varicose." It is that which has been found connected with *arterial incompetency*, and which I propose therefore to designate the *arterial* ulcer. It is a very characteristic sore, usually situated in the midst of a portion of dense tegument, of a pale pinkish hue. Its size varies from that of a small and somewhat superficial ulcer with a welted edge—as it is often seen, creeping along the leg behind the inner malleolus and tibia,—to an ulcer pervading the whole leg, excepting a few small islands of *dense*, thickened integument, and making its way gradually through the aponeurosis into the deeper structures. Such an ulcer results from impairment of nutrition, through defective supply of arterial blood; and the *welt*, as well as the indurated integument, shows various degrees of such impairment short of *vital extinction*. The welt commonly seen in the *least* severe cases of this class is, in all probability, due to nutritive efforts almost at the *confines* of the *arterial* circulation. It is the ulcer with "defect of action" of Mr. Syme. I need not say that this is a still more formidable variety of ulcer, and is, I believe, incurable by any remedies that have hitherto been devised for it.

To recapitulate. I have attempted to show,—

1st. That there are no substantial grounds for accrediting ulcers, on legs *with* varicose veins, to the diseased veins, in the relation of effect; that, in fact, the varicose ulcer, in the sense in which it is usually understood, is a fiction.

2nd. That when ulcer and varicosity co-exist the veins are much more likely to suffer injury from the ulcer, than the ulcer from the veins. In fact, the veins cannot be included within the confines of the ulcerative process without suffering in common with the other ulcer tissues.

3rd. That ulcers of the leg are, in relation to their pathological genesis, divisible into three species :—

1. The *simple* ulcer; that in which the morbid processes are limited to the confines of the tissues directly engaged therein; 2, the *venous* ulcer; most frequently with "bronzing" and induration of the skin; dependent upon obstructive disease of the trunk veins, superficial or deep; and, 3, the *arterial* ulcer, generally without bronzing, but with induration of the tegument, and due to incompetency of the arteries, through disease of their coats. In some cases, however, the morbid changes which characterize the second and third classes are conjoined.

All of these ulcers are capable of being more or less varied by secondary agencies; and thus subordinate varieties are determined.

By way of conclusion, I must make a few remarks on the subject of treatment. This, however, becomes moderately simple, if the selection of remedies is based upon a recognition of the pathological changes with which these varied forms of ulcer are respectively associated, and of the agencies by which they became *secondarily* modified.

There is one rule which is applicable to all leg ulcers—namely, that *rest* is an essential part of their treatment,—with the foot elevated above the pelvis in those of the *first* and *second* classes, and on a level with it, or inclined in the opposite direction, in those of the *third* class.

First, with respect to the *simple* ulcer. As I have said, there are no peculiarities in its condition that cannot be reconciled with *action* in its immediate confines, modified by constitutional derangement. If acutely inflamed, bleeding is required by leeches or scarification across the edge of the ulcer, hot applications, constitutional treatment, purgatives, calomel when the biliary system is deranged, and opiates if the sore is painful. If the inflammation is only subacute, and especially if the tissues be thickened and the edge welted, blisters should be applied over the surface and the affected border. It may safely be alleged that no one remedy is capable of so much general benefit as the blister, for which we are, I believe, indebted to M. Sappey, but mainly to Mr. Syme. Irritability must be subdued by opium, with quinine or iron, and the use of a *strong* solution of nitrate of silver to the surface. If neuralgia exist, the offending nerve-filament must be tracked out and divided, as Mr. Hilton recommends, and appropriate constitutional remedies administered. If there be constitutional weakness, tonics and change of air, with generous diet, are required; and if struma or syphilis, the remedies appropriate to these diseases. If the ulcer persists after the subjugation of secondary states or conditions and a certain period of rest, the *surface* should be destroyed by some powerfully destructive agent. The strong solution of the pernitrate of mercury, twice or thrice applied, and at intervals of two or three days, will seldom fail to induce cicatrization in such a sore. And here I may remark of this remedy that, when applied to certain specific sores, such as syphilitic lupus, and especially secondary syphilitic *ulcers about the throat and palate*, it seldom fails *at once* to change their character. To these latter parts, however, it must be very carefully applied. In tedious and doubtful cases of simple chronic ulcer I have seen the best results follow the administration of Liqr. Hyd. Bichlorid. with Potas. Iodid.

With respect to the *venous* ulcer, I know but of one mode of inducing permanent cicatrization—namely, by those incisions at the edges to which I first drew the attention of the profession through the medium of this Society, in 1861. I must remind you of its pathology, for upon this all valid treatment must be based— viz., condensation and contraction of the superfascial textures, obliteration or destruction of the superficial veins, sometimes with obstruction of their trunks and concurrent disease of the deep arteries and veins, and the establishment of collateral but imperfect venous circulation.

No palliative treatment can be of any permanent benefit to such a sore, although it may be from time to time required to meet contingent disorders, such as attacks of inflammation, pain in the ulcer surface, hæmorrhage, œdema, &c., &c. For ulcers of this class bandages are not generally indicated.

There is, however, one kind of *pain* which often accompanies a sore which spreads in the direction of and over the anterior surface of the tibia. It is severe and chiefly nocturnal, arises from periosteal complication, and requires opium, iodide of potash, and occasionally free subcutaneous division of the periosteum for its relief.

The radical cure of these ulcers contemplates their *permanent cicatrization;* for which the proposal simply to tie the veins and cut away the edges is unavailing.

The *principle* of treatment involves the destruction of any veins which pass from the margin of the diseased skin tissue, and especially from the ulcer itself; and relief to the tension of the contracted and hardened skin. These two objects are to be carried out at the same time; 1st, by deligation of the larger varicose veins; and 2nd, by curved incisions, free and deep, on either side of the ulcer margin (Plate V., fig. 1, x, x). The blood will flow freely from the veins which will thus be divided; but this can be very readily controlled by elevating the leg, plugging, and the temporary application of a firm bandage. The edges of the wound should be kept apart, and the limb maintained most scrupulously, and without intermission, in the posture described when speaking of obliteration of veins.

I trust you will not think that I am over-indulgent to my own suggestion, or desirous of estimating it above its real worth. I have now tried these incisions in a large number of cases, and so have other surgeons, and I have had every reason to be satisfied with the results. Ulcers of from one to twenty years' standing have thus been induced to heal. It may be necessary to repeat the incisions, sometimes even to two or three, on either side of the ulcer. I have noted repeatedly, by measuring the limb, that the distance between the edges of the cicatrices has gradually increased; and that the dimensions of a before-contracted limb have thereby increased. The cicatrix has uniformly remained firm, although, as in two or three cases, a fresh ulcer has formed in close contiguity to it after a lapse of years.

Whatever the *condition* of the diseased skin tissue, whether firm like leather, or spongy, answering to the "tissue fungueux or caverneux" of the French writers, a wound made in it *heals*, and sometimes with great rapidity. It not unfrequently happens that the ulcer heals first, in consequence of some little sloughing on the surface of the artificial wounds. This is, however, a matter of no consequence.

The *arterial* ulcer is incurable by any known means of which I am cognizant; and amputation is a resort full of risk, as the ulcer is so commonly associated with remote organic, especially renal, disease. Would ligature or compression of the femoral artery, with the object of closing up the diseased arterial trunks, and of obtaining a renovated arterial system by means of those abundant new branches and their ample anastomosis, which was so striking a feature in the injected case of which I have spoken, be likely to meet the otherwise insuperable difficulty?

I have now completed these Lectures, and have to thank my audience for their indulgent, but, I fear, unrequited attention.

www.ingramcontent.com/pod-product-compliance
Lightning Source LLC
Chambersburg PA
CBHW020240170426
43202CB00008B/166